Lecture Notes in Mathematics

Edited by A. Dold and B. Eckmann

696

Philip J. Feinsilver

Special Functions, Probability Semigroups, and Hamiltonian Flows

Springer-Verlag
Berlin Heidelberg New York 1978

Author
Philip J. Feinsilver
Department of Mathematics
Southern Illinois University
Carbondale, Il 62901/USA

AMS Subject Classifications (1970): 33A30, 33A65, 39A15, 42A52, 44A45, 47D10, 60H05, 60J35, 81A20

ISBN 3-540-09100-9 Springer-Verlag Berlin Heidelberg New York
ISBN 0-387-09100-9 Springer-Verlag New York Heidelberg Berlin

Printed in Germany

Printing and binding: Beltz Offsetdruck, Hemsbach/Bergstr.
2141/3140-543210

Summary of Contents.

Chapter I.

We introduce the generator corresponding to a process with independent increments. Assuming we have a convolution semigroup of measures satisfying $\int e^{zx}p_t(dx) = e^{tL(z)}$ where $L(z)$ is analytic in a neighborhood of $0\in\mathbf{C}$, only special L's arise. Conversely, for any $L(z)$ analytic near $0\in\mathbf{C}$, $L(0) = 0$, define a general translation-invariant process to have corresponding densities $p_t(x) = \frac{1}{2\pi}\int e^{-i\xi x}e^{tL(i\xi)}d\xi$; $p_t(x)$ may not be positive measures. Our basic tools to analyze the corresponding processes $w(t)$ such that $\frac{1}{t}\log\langle e^{zw(t)}\rangle = L(z)$, $\langle\rangle$ denoting expected value are:

(1) Hamiltonian flows, e.g. choosing $H = L(z)$, z = momentum.

(2) Iterated stochastic integrals.

Chapter II.

We study the basic theory of operators analytic in (x,D), $D = \frac{d}{dx}$ and of functions $f(x)$, $f(D)$, $f\in\mathcal{S}$ (Schwartz space) or $f\in\mathcal{S}^*$.

Chapter III.

We study the generalized powers $x(t)^n$ where $x(t) = e^{tH}xe^{-tH}$, H a Hamiltonian such that $H1 = 0$. We introduce the operator $A = x(t)z(t)$, a generalization of xD.

Chapter IV.

We study orthogonal polynomials corresponding to a certain class of generators, which we call Bernoulli generators. The main feature is that the orthogonal systems are actually generalized powers and so the orthogonal series' are isomorphic to Taylor series'.

Chapter V.

We study in detail the five standard Bernoulli-type processes: Bernoulli, Symmetric Bernoulli, Exponential, Poisson, Brownian Motion. The most familiar special functions appear. In general we see that the relevant functions are confluent hypergeometric functions.

Chapter VI.

We discuss the relationships among the five standard processes. We have

$$\begin{array}{ccccc} & \overset{Q}{\nearrow} & \text{Exponential, } e(t) & \overset{r^{-1}}{\searrow} & \\ \begin{array}{c}\text{Bernoulli}\\ x(t)\end{array} & \xrightarrow{\alpha} & \begin{array}{c}\text{Symmetric}\\ \text{Bernoulli}\end{array}\ d(t) & \xrightarrow{\beta} & \begin{array}{c}\text{Brownian Motion}\\ b(t)\end{array} \\ & \underset{\beta}{\searrow} & \text{Poisson, } N(t) & \underset{\alpha}{\nearrow} & \end{array}$$

The arrows indicate limits taken by the vanishing of the indicated parameter (these parameters determine the various generators L).

Chapter VII.

We discuss the theory of discrete iterated integrals (sums) and some corresponding limits, thus determining iterated stochastic and deterministic integrals.

Chapter VIII.

We study the theory of Chapters III & IV extended to vector processes $\underset{\sim}{w}(t) \in R^N$.

Chapter IX.

We clarify the correspondence between the (Markov) stochastic process with generator H(x,D) and the quantum process with Hamiltonian H(x,z).

Chapter X.

We discuss lines for further developments.

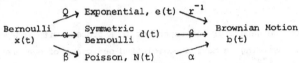

TABLE OF CONTENTS

CHAPTER I. INTRODUCTION

The probabilistic genesis of the topics we will discuss is found in considering limiting distributions for sums of independent random variables. We will find that the processes considered are limiting cases of the model of sequential independent coin flips, where we define, in the discrete case, S_n = # of heads in n tosses. Write, more generally, $S_n = X_1 + X_2 + \ldots + X_n$, where X_j are independent and identically distributed, with density $p(x)$. Then, for $z \in \mathbb{C}$, Re $z = 0$, $\langle\ \rangle$ denoting expected value,

$$\langle e^{zS_n}\rangle = \left(\int e^{zx}p(x)\right)^n = \int e^{zx}p_n(x) \qquad \text{where}$$

p_n is the density for S_n. We assume that

$$L(z) = \log\langle e^{zX}\rangle = \log\int e^{zx}p(dx)$$

can be extended to an analytic function in a neighborhood of $0 \in \mathbb{C}$. Passing to continuous time we have a convolution semigroup of measures $p_t(x)$ [or $p_t(dx)$] such that

$$e^{tL(z)} = \int e^{zx}p_t(x).$$

The corresponding process we denote by $w(t)$, i.e., $w(t)$ is a process with stationary independent increments and

$$\langle e^{zw(t)}\rangle = e^{tL(z)}.$$

Given $L(z)$, analytic near 0, we can define

$$p_t(x) = \frac{1}{2\pi}\int e^{-i\xi x}e^{tL(i\xi)}d\xi.$$

$p_t(x)$ need not be a positive measure; however, we will assume that $L(0) = 0$ so that $\int p_t = 1$. $L(z)$ is called the generator.

In the following we will see that many of the "special functions" familiar from physics appear as canonical constructs in our theory. Usually they appear as

(1) generalized powers x^n

(2) reproducing kernels for the space of functionals associated with L

(3) types of generating functions or densities.

The functions that arise will satisfy equations of the type $\frac{\partial^2 u}{\partial t^2} = H^2u$. On the space of paths $w(t)$, generated by H, then, $u[w(t),t]$ or $u[w(t),-t]$ will be a martingale.

Probabilistically, a basis for functionals of $w(t)$ is most conveniently sought by iterated non-anticipating stochastic integration. This construction is explained in terms of discretization, replacing integrals by sums, in Chapter VII. Another probabilistic aspect of the theory will be classes of limit theorems:

(1) Limit theorems among our standard processes (see Chapter VI)

(2) Limit theorems for symmetric functionals via passage from the discrete integrals to the continuous (Chapter VII).

Guiding us are two major principles. The first is that the construction "iterated non-anticipating integrals" is at the heart of the various types of orthogonal and Taylor expansions. We will see, e.g., that even theta functions arise via these constructions, as "exponentials."

The second principle is that the generator L should be thought of as the Hamiltonian of a quantum dynamical system. We will use this principle both as a motivational and computational guide throughout.

Now let's proceed to the detailed exposition.

PART I. ANALYTICAL AND PROBABILISTIC BASIS

CHAPTER II. BASIC OPERATOR THEORY

We will deal with functions of a real (or complex) variable x and of $D = \frac{d}{dx}$. Thinking of a function f(x) as a multiplication operator, we "recover" the function as f(x)1. We will denote, then, for an operator B, the operator composition by $B^\circ f(x)$; and the application of B to f(x) by Bf(x), that is, $Bf(x) = B^\circ f(x)1$.

The functional calculus we use will be based on the exponential. We have the following:

<u>Proposition 1</u>: Knowledge of e^{aB} is equivalent to knowledge of f(B) for the following families {f}:

(1) Polynomials

(2) Analytic functions (around $0 \in \mathbb{C}$)

(3) Schwartz space functions and tempered distributions.

<u>Proof</u>: For (1) calculate $B^n = (\frac{d}{da})^n\big|_0 e^{aB}$. (2) follows by power series expansion. From (1) or (2) we recover e^{aB} by power series.

Recall Schwartz space $S = \{f : f \in C^\infty$ and $\lim_{|x|\to\infty} |x|^n D^m f(x) = 0$, for all $n,m \geq 0\}$. $S^* = \{$tempered distributions$\}$. Then, for example,

$$\delta(B) = \frac{1}{2\pi}\int e^{iyB}dy$$

and generally for $f \in S$ or S^*,

$$f(B) = \int e^{iyB}\hat{f}(y)dy.$$

<u>Remark</u>: We use the normalization $\hat{f}(y) = \frac{1}{2\pi}\int e^{-iyx}f(x)dx$.

As a runs from $-\infty$ to ∞ e^{aB} forms a group of operators. We are tacitly assuming that a suitable domain exists. A basic means of computing, or defining, $e^{aB}f$ is as the solution u to

$$\frac{\partial u}{\partial a} = Bu , \quad u(0) = f.$$

The observation that enables us to use a quantum-mechanical viewpoint is simply this. Assume that B1 = 0. Then if the operator U satisfies

$$\frac{\partial U}{\partial a} = [B,U] = BU - UB , \quad U(0) = f,$$

we see that $U = e^{aB}fe^{-aB}$ and u = U1. So we can always consider the evolution equations determining exponentials as operator equations. Using exponentials as a basis for our functional calculus we can determine inverse operators too.

We define

$$B^{-1} = \int_0^1 \lambda^B \frac{d\lambda}{\lambda} = \int_0^\infty e^{-yB}dy$$

and the generating function (resolvent) $\sum_0^\infty \frac{z^n}{B^{n+1}}$ by

$$\frac{1}{B-z} = \int_0^\infty e^{zy}e^{-yB}dy.$$

We will denote the Heaviside function by χ. Thus,

$$\chi(x) = D^{-1}\delta(x) = \frac{1}{2\pi}\int \frac{e^{iyx}}{iy}\,dy = \int_0^\infty e^{-yD}\delta(x)dy.$$

HEISENBERG GROUP FORMULATION

Given any two operators R,S such that $[R,S] = 1$, and $R1 = 0$, we can establish a calculus. For clarity we denote our operators by D and x, noting that $R \leftrightarrow D$, $S \leftrightarrow x$ establishes an isomorphism of the given (R,S) system and the familiar one. Our first theorem is the

Generalized Leibniz Lemma (GLM)

$$g(D)^\circ f(x) = \sum_0^\infty \frac{f^{(n)}(x)g^{(n)}(D)}{n!}$$

Remark: This allows us to express all products with derivative operators on the right.

Proof:

Step 1. $D^n\,{}^\circ x = xD^n + nD^{n-1}$, $n > 0$.

$n = 1$: Definition $[D,x] = 1$.

$n = m+1$: Multiply $D^m\,{}^\circ x = xD^m + mD^{m-1}$ on the left by D.

Then $D^{m+1}\,{}^\circ x = xD^{m+1} + (m+1)D^m$ follows.

Step 2. Multiplying $D^n\,{}^\circ x = xD^n + nD^{n-1}$ by $\frac{t^n}{n!}$ and summing yields

$$e^{tD}\,{}^\circ x = xe^{tD} + te^{tD} = (x+t)e^{tD}.$$

Induction immediately yields

$$e^{tD}\,{}^\circ x^n = (x+t)^n e^{tD}$$

and hence

$$e^{tD}\,{}^\circ e^{sx} = e^{sx}e^{st}e^{tD}.$$

Step 3. Therefore

$$e^{tD}\,{}^\circ e^{sx} = \sum_0^\infty \frac{1}{n!}\, s^n\, e^{sx} t^n e^{tD} = \sum_0^\infty \frac{1}{n!}\, \partial_x^n e^{sx} \partial_D^n e^{tD}$$

Applying Proposition 1 yields the result.

<u>Notational Remark</u>: ∂_X denotes differentiation with respect to the argument X.

<u>Corollary</u>: $[g(D),x] = g'(D)$. $[D,f(x)] = f'(x)$.

<u>Proposition 2</u>: Given an operator V(D) we can define a canonical dual $\xi = xW$, where $W = \frac{1}{V'(D)}$, such that $[V,\xi] = 1$.

<u>Proof</u>: By GLM, $VxW = xVW + V'W = xVW + 1$.

From the proof of GLM we have the

<u>Exponential Lemma (EL)</u>

$$e^{tD}f(x) = f(x+t) \quad \text{and} \quad g(D)e^{sx} = g(s)e^{sx}.$$

<u>Proof</u>: $e^{tD}e^{sx} = e^{tD}\, {}^{o}e^{sx}1 = e^{sx}e^{st}e^{tD}1 = e^{sx}e^{st}$, since $D1 = 0$.

Apply Proposition 1, writing first

$$e^{tD}e^{sx} = e^{s(x+t)} \text{ to yield } e^{tD}f(x) = f(x+t)$$

and then

$$e^{tD}e^{sx} = e^{ts}e^{sx} \text{ for the second result.}$$

<u>Proposition 3</u>: Let V and ξ be canonical duals and assume that V(D) is an invertible function with inverse U. Then

(1) $e^{aV}f(\xi)1 = f(\xi+a)1$

(2) $e^{a\xi}e^{bx} = e^{xU[a+V(b)]}$

<u>Proof</u>: (1) is just the exponential lemma.

(2) $F = e^{a\xi}e^{bx}$ is the solution to $\frac{\partial F}{\partial a} = \xi F$, $F(0) = e^{bx}$.

Denote $E(a,b) = e^{xU[a+V(b)]}$. Observe that $E(0,b) = e^{bx}$

and $\frac{\partial E}{\partial a} = x\partial_a U[a+V(b)]E$

and that

$$WE = \frac{1}{V'\{U[a+V(b)]\}} E \text{ by the EL.}$$

We have our third basic

<u>Duality Lemma</u>

$$g(D)f(x) = f(\partial_a)e^{ax}g(a)\Big|_{a=0}.$$

<u>Proof</u>: To apply Proposition 1 we check, using EL:

$$e^{tD}f(x) = f(x+t) = f(\partial_a)e^{a(x+t)}\Big|_0 = f(\partial_a)e^{ax}e^{ta}\Big|_0.$$

A simple corollary is

$$g(D)1 = e^{ax}g(a)\Big|_0 = g(0).$$

THE OPERATOR xD

The operator xD has eigenfunctions x^n and since $xDx^n = nx^n$ it is our prototype "number operator." It also acts something like a logarithm, since $\log_x x^n = n$; and it is easy to see by the GLM that it is the canonical dual of the singular operator logD.

Proposition 4: $\lambda^{xD}f(x) = f(x\lambda)$.

Proof: Set $\lambda = e^s$. Check that $u = f(xe^s)$ satisfies

$$\frac{\partial u}{\partial s} = xe^s f'(xe^s) = xDu\ , \quad u(0) = f(x).$$

Example: It is immediate to generalize to N dimensions, defining $A = x\cdot D$ where D = gradient operator. Then if f is homogeneous of degree d,

$$\lambda^A f(x) = f(x\lambda) = \lambda^d f(x)$$

and Euler's theorem follows by applying ∂_λ at $\lambda = 1$.

HAMILTONIAN FORMULATION

In the following we will use the Hamiltonian method for determining exponentials of operators. We use the notations z and D interchangeably, thinking of z as "momentum." z and x are Heisenberg duals, $z1 = 0$.

Proposition 5: Given a Hamiltonian $H(x,z)$ (assumed analytic in both variables). Define $z(0) = z$, $x(0) = x$, $z(t) = e^{tH}ze^{-tH}$, $x(t) = e^{tH}xe^{-tH}$.

Then (1) $[z(t),x(t)] = 1$

$$(2)\ \dot{x}(t) = [H,x(t)] = \frac{\partial H}{\partial z}[x(t),z(t)]$$

$$(3)\ \dot{z}(t) = [H,z(t)] = -\frac{\partial H}{\partial x}[x(t),z(t)]$$

where the dot denotes $\frac{d}{dt}$.

Proof: (1) follows from $[z,x] = 1$.

We can assume H has an expansion with all z operators on the right.

$$\dot{x} = [H,x(t)] = [H,e^{tH}xe^{-tH}] = e^{tH}[H,x]e^{-tH}$$

$$= e^{tH}\frac{\partial H}{\partial z}e^{-tH} = \frac{\partial H}{\partial z}[x(t),z(t)], \text{ by GLM.}$$

Similarly

$$\dot{z} = [H,z(t)] = e^{tH}[H,z]e^{-tH} = -\frac{\partial H}{\partial x}[x(t),z(t)].$$

Corollary: If $H1 = 0$, then $e^{tH}f(x) = f[x(t)]1$.

Remark: We will always use (or assume) H's such that $H1 = 0$.

SOME CONTOURS

We conclude this chapter with a discussion of some basic contours to be used for integral representations of functions and operators.

(1) The contour O denotes a small circle around 0 in **C**. We can, of course, represent

$$\frac{1}{\Gamma(n)} = \frac{1}{2\pi i}\int_O e^y y^{-n}\,dy, \quad n \geq 0.$$

As another example,

$$B^n = \frac{n!}{2\pi i}\int_O \frac{e^{yB}}{y^{n+1}}\,dy.$$

(2) In order to get $\Gamma(n+t)$ for $n+t$ not an integer we use the contour H (named after Hankel) that avoids the branch point at 0, by going from $-\infty$ around 0 and back to $-\infty$ along the negative real axis. Thus, for $t > 0$,

$$\frac{1}{\Gamma(t)} = \frac{1}{2\pi i}\int_H e^y y^{-t}\,dy$$

and
$$B^{t-1} = \frac{\Gamma(t)}{2\pi i}\int_H e^{yB} y^{-t}\,dy.$$

Before proceeding further we make

Notational Remark: We use

$$t_{(n)} = \frac{\Gamma(t+n)}{\Gamma(t)} = t(t+1)\ldots(t+n-1)$$

and

$$t^{(n)} = (-1)^n(-t)_{(n)} = t(t-1)\ldots(t-n+1) = n^{\text{th}} \text{ factorial power of } t.$$

E.g., $(1+x)^n = \Sigma \frac{n^{(k)}}{k!} x^k$.

Proposition 6: Let f have an expansion $\sum_0^\infty \frac{x^n}{n!} f_n$.

Define $F_t(x) = \sum_0^\infty \frac{x^n f_n}{n!\,t_{(n)}}$

Then $F_t(x) = \frac{\Gamma(t)}{2\pi i}\int_H e^y y^{-t} f(\frac{x}{y})\,dy$

Proof:
$$F_t(x) = \Sigma \frac{x^n f_n}{n!\,t_{(n)}} = \Sigma \frac{x^n f_n}{n!}\frac{\Gamma(t)}{\Gamma(t+n)} = \frac{\Gamma(t)}{2\pi i}\int_H \Sigma \frac{x^n f_n}{n!} e^y y^{-n-t}\,dy,$$

now resum.

CHAPTER III. MOMENT THEORY

Consider now a convolution semigroup $p_t(x)$ and corresponding process $w(t)$ such that

$$\int e^{ax}p_t(x) = \langle e^{aw(t)}\rangle = e^{tL(a)}.$$

We will assume L is analytic around 0 and $L(0) = 0$.

Remark: Substituting an operator B for "a" in the above yields the exponential for the operator $L(B)$.

Proposition 7: (1) $e^{tL(D)}f(x) = \langle f[x+w(t)]\rangle$

(2) $p_t(y-x) = e^{tL(D)}\delta(x-y)$, i.e. $p_t(y-x)$ is the fundamental solution to $\frac{\partial u}{\partial t} = Lu$.

(3) $e^{tL(D)}f(x) = \int f(x+y)p_t(y)$

Proof: (1) $e^{tL(D)}f(x) = \langle e^{Dw(t)}f(x)\rangle = \langle f[x+w(t)]\rangle$ by EL.

(2) $e^{tL(D)}\delta(x-y) = \int e^{sD}p_t(s)\delta(x-y) = \int \delta(x+s-y)p_t(s) = p_t(y-x)$.

(3) Apply $\int f(y)dy$ to (2).

THE OPERATOR C

We now consider the flow generated by L. We have:

$$H = L(z)$$
$$\dot{x}(t) = L'[z(t)]$$
$$\dot{z}(t) = 0.$$

So $z(t) = z$ and $\dot{x}(t) = L'(z)$, yielding

$$x(t) = x + tL'(z).$$

This operator, $x(t)$, we denote simply by C. Thus,

$$\langle f[x+w(t)]\rangle = e^{tL}f(x) = f[x(t)]1 = f(C)1.$$

Proposition 8: $e^{aC}f(x) = e^{ax}e^{t[L(D+a)-L(D)]}f(x)$

Proof: $e^{aC}f(x) = e^{ax(t)}f(x) = e^{tL}e^{ax}e^{-tL}f(x)$.

Apply Proposition 1 to

$$e^{tL}e^{ax}e^{-tL}e^{bx} = e^{tL}e^{ax}e^{bx-tL(b)} = e^{ax}e^{t[L(a+b)-L(b)]}e^{bx}$$
$$= e^{ax}e^{t[L(a+D)-L(D)]}e^{bx} \text{ by EL.}$$

MOMENT POLYNOMIALS

We will use the notation $\mu_k(t)$ to denote the moments of $p_t(x)$. That is,

$$\mu_k(t) = \int y^k p_t(y) = \partial_a^k\Big|_{a=0} e^{tL(a)}.$$

So, for example,

$$e^{tL(a)} = \sum_0^\infty \frac{a^n}{n!}\mu_n(t).$$

Definition: The moment polynomials associated with L are the polynomials $h_n(x,t) = x(t)^n 1$, i.e. $C^n 1$.

Proposition 9: Observe that

(1) $h_n(x,t) = e^{tL}x^n$

(2) $h_n(x,t) = \Sigma\binom{n}{k}x^{n-k}\mu_k(t)$

Proof: (1) $C^n 1 = e^{tL}x^n e^{-tL}1 = e^{tL}x^n$.

(2) The duality lemma implies

$$e^{tL}x^n = (\partial_a)^n e^{ax}e^{tL(a)}\Big|_{a=0} = \Sigma\binom{n}{k}x^{n-k}\mu_k(t)$$

by the usual Leibniz rule.

We define the generating function for the h_n,

$$g(x,t;a) = \Sigma\frac{a^n}{n!}h_n.$$

It is easy to see that

$$g(x,t;a) = e^{aC}1 = e^{tL}e^{ax}e^{-tL}1 = e^{ax+tL(a)}.$$

Proposition 10: h_n are generalized powers, satisfying

(1) $h_o = 1$

(2) $Ch_n = h_{n+1}$

(3) $Dh_n = nh_{n-1}$

(4) $\frac{\partial h_n}{\partial t} = Lh_n$

(5) $CDh_n = nh_n$

Proof: (1) and (2) are definitions.

$Dh_n = z(t)x(t)^n 1 = x(t)^n z(t)1 + nx(t)^{n-1}1$ by Proposition 5 and GLM.

Since $z(t)1 = z1 = 0$, (3) follows.

(4) follows from $h_n = e^{tL}x^n$.

(5) follows from (2) and (3).

Notation: We will denote the operator CD by A.

Remarks: (1) For any Hamiltonian H, the operator $A = x(t)z(t)$ satisfies

$Ax(t)^n 1 = nx(t)^n 1$ since

$$Ax(t)^n 1 = e^{tH} xzx^n e^{-tH} 1 = nx(t)^n 1 + e^{tH} x^{n+1} z e^{-tH} 1.$$

So, $Ax(t)^n 1 = nx(t)^n 1$.

(2) We define the conjugate operator $\bar{C}$ and the conjugate system $\bar{h}_n(x,t)$ by substituting $t \to -t$.

In general, we use the bar to denote an operator or function transformed by changing t to -t.

FUNCTIONS WITH NEGATIVE INDEX

We want to define $h_{-(n+1)} = C^{-(n+1)} 1$, $n > 0$. As indicated in Chapter II, we define $g^-(x,t;a)$ by

$$\Sigma a^n h_{-(n+1)} = \Sigma \frac{a^n}{C^{n+1}} 1 = \int_0^\infty e^{ay} e^{-yC} 1 dy = \int_0^\infty e^{y(a-x)} e^{tL(-y)} dy.$$

Remark: The integral may require a change $t \to -t$ or may diverge regardless. These h's have properties similar to h_n's and are generalized negative powers. (See last section of this chapter.)

VACUUM FUNCTIONS

In Proposition 10 we noted that starting with $h_0 = 1$, $C^n h_0$ satisfy the two equations (4) and (5) that are immediate for $h_0 = 1$. We ask what functions Ω have the property that for $\Omega_n = C^n \Omega$

(1) $\dfrac{\partial \Omega_n}{\partial t} = L\Omega_n$

(2) $A\Omega_n = n\Omega_n$ for all $n \geq 0$.

Definition: Given a Hamiltonian H, a function u is defined to be:

harmonic if $\dfrac{\partial u}{\partial t} + Hu = 0$ and

coharmonic if $\dfrac{\partial u}{\partial t} = Hu$ (since then $\bar{u}$ is harmonic).

Similarly an operator U is harmonic if it satisfies $\dfrac{\partial U}{\partial t} + [H,U] = 0$ or coharmonic if $\bar{U}$ is harmonic.

Remarks: (1) A harmonic function evaluated along the flow induced by H is independent of time. That is,

$u[x(t),t]1$ is independent of t.

We check this:

$u[x(t),t]1 = e^{tH} u(x,t)$ implies

$$\frac{du}{dt}[x(t),t]1 = e^{tH} Hu(x,t) + e^{tH} \frac{\partial u}{\partial t} = 0.$$

Alternatively, $e^{tH} u(x,t) = e^{tH} e^{-tH} u(x,0) = u(x,0)$.

(2) Similarly a harmonic operator U is a "constant of the motion" generated by H. Check as above by differentiating, or using

$$U = e^{-tH}U(0)e^{tH} \text{ to get}$$

$$e^{tH}Ue^{-tH} = e^{tH}e^{-tH}U(0)e^{tH}e^{-tH} = U(0).$$

(3) On the path space generated by H, a harmonic u yields a martingale $u[w(t),t]$.

(4) If u is either harmonic or coharmonic, we have $\frac{\partial^2 u}{\partial t^2} = H^2 u$.

So we are seeking Ω's that preserve coharmonicity under the action of C and that yield eigenfunctions of A. We will consider general H [not necessarily just $H = L(z)$].

<u>Definition</u>: A function Ω is a vacuum function if it satisfies

(1) Ω is coharmonic.

(2) $A\Omega = x(t)z(t)\Omega = 0$.

Ω is an <u>absolute</u> <u>vacuum</u> if, in addition, $\frac{\partial\Omega}{\partial t} = 0$.

<u>Remark</u>: If Ω is absolute, conditions (1) and (2) become

(1) $H\Omega = 0$

(2) $xz\Omega = 0$.

We have the following:

<u>Theorem 1</u> Ω is a vacuum function if and only if $\Omega_n = x(t)^n\Omega$ satisfy:

(1) Ω_n is coharmonic for each n.

(2) $A\Omega_n = n\Omega_n$

<u>Proof</u>: Ω is a vacuum implies that $\Omega = e^{tH}\Omega(0)$.

Thus,

$$x(t)^n\Omega = e^{tH}x^n e^{-tH}e^{tH}\Omega(0) = e^{tH}x^n\Omega(0) \text{ is coharmonic.}$$

And,

$$\begin{aligned} A\Omega_n &= x(t)z(t)x(t)^n\Omega = e^{tH}xzx^n e^{-tH}\Omega \\ &= nx(t)^n\Omega + e^{tH}x^{n+1}ze^{-tH}\Omega \\ &= n\Omega_n + x(t)^n x(t)z(t)\Omega \\ &= n\Omega_n. \end{aligned}$$

The converse follows by setting $n = 0$.

CANONICAL VACUUMS FOR $H = L(z)$

There are two "natural" choices for the vacuum. Observe that $xDh_o = h_o$ has the solution 1 and also the distribution solution $\chi(x)$. We thus have

Proposition 11: $\Omega = D^{-1}p_t(-x) = e^{tL}\chi(x) = \chi(C)1$ is a vacuum function.

Proof: $e^{tL}\chi$ is coharmonic by definition.
Proposition 7(2) implies

$$e^{tL}\chi(x) = e^{tL}D^{-1}\delta(x) = D^{-1}e^{tL}\delta(x) = D^{-1}p_t(-x).$$

Also, $A\Omega = e^{tL}xDe^{-tL}e^{tL}D^{-1}\delta(x) = e^{tL}x\delta(x) = 0.$

Notation: Given L, then, Ω will denote this canonical vacuum.

We thus have two more series starting with Ω.

Definition:

$$p_n(x,t) = C^n\Omega.$$

$$p_{-(n+1)}(x,t) = \frac{(-1)^n}{n!}D^{n+1}\Omega.$$

We denote $g_+ = e^{aC}\Omega = e^{aC}\chi(C)1$, since $g_+(x,0;a) = e^{ax}\chi(x)$.

The series for negative index is derived as follows. We want $p_{-(n+1)}$ to satisfy

(1) $Cp_{-(n+1)} = p_{-n}$

(2) $Dp_{-(n+1)} = -(n+1)p_{-(n+2)}$

(3) $CDp_{-(n+1)} = -(n+1)p_{-(n+1)}$.

Starting with $p_0 = \Omega$, we use property (2) to generate the p's. Note that this procedure would just give 0 using 1 as the vacuum. We thus have $p_{-1} = p_t(-x)$.

Proposition 12: $p_{-(n+1)} = \frac{(-1)^n D^n}{n!} p_t(-x)$ and satisfies properties (1)-(3) above, for $n > 0$.

Proof: We remark that there would be difficulty at $n = 0$, since

$$Cp_{-1} = Cp_t(-x) = e^{tL}x\delta(x) = 0;$$

which is why the series starts at p_{-1}. And

$$Cp_{-(n+1)} = \frac{(-1)^n}{n!}[D^nCp_t(-x) - nD^{n-1}p_t(-x)] = \frac{(-1)^{n-1}}{(n-1)!}D^{n-1}p_t(-x)$$

follows using $Cp_{-1} = 0$.

We denote $g_+^- = \Sigma a^n p_{-(n+1)} = \Sigma \frac{a^n(-1)^nD^n}{n!} p_t(-x) = p_t(a-x)$, and we check

$$g_+^- = \Sigma \frac{a^n(-1)^n}{n!} D^{n+1}\chi(C)1 = D\chi(C-a)1 = e^{tL}D\chi(x-a) = p_t(a-x).$$

We can relate pairs of g's as follows.

Proposition 13: (1) $g^-(x,t;a) = \int_0^\infty e^{ay} g(x,t;-y)dy$

(2) $g_+(x,t;a) = \int_0^\infty e^{ay} g_+^-(x,t;y)dy$

Proof: (1): Substitute in $\int_0^\infty e^{ya} e^{-yx+tL(-y)} dy$.

(2): Note that since, for test functions φ,

$$\int \varphi(x) \int_0^\infty \delta(x-y) f(y)dy = \int_0^\infty \varphi(y) f(y) = \int \varphi(x) \chi(x) f(x),$$

we have

$$\int_0^\infty \delta(x-y) f(y)dy = \chi(x) f(x).$$

Thus,

$$g_+ = e^{tL} e^{ax} \chi(x) = e^{tL} \int_0^\infty \delta(x-y) e^{ay} dy = \int_0^\infty e^{ay} p_t(y-x)dy.$$

DIFFERENTIAL AND INTEGRAL REPRESENTATIONS

We recall the formula from Chapter II, $B^n = \frac{n!}{2\pi i} \int_0 e^{yB} y^{-(n+1)} dy$. Apply this, with C instead of B, to 1 and Ω to get:

(1) $h_n = C^n 1 = \frac{n!}{2\pi i} \int_0 g(x,t;y) y^{-(n+1)} dy = \frac{\partial^n}{\partial y^n} \Big|_{y=0} g(x,t;y),$

(2) $p_n = C^n \Omega = \frac{n!}{2\pi i} \int_0 g_+(x,t;y) y^{-(n+1)} dy = \frac{\partial^n}{\partial y^n} \Big|_{y=0} g_+(x,t;y).$

From $g^-(x,t;y) = \int_0^\infty e^{s(y-x)} e^{tL(-s)} ds$, we get

(3) $h_{-(n+1)} = \frac{1}{n!} \frac{\partial^n}{\partial y^n}\Big|_{y=0} g^-(x,t;y) = \frac{1}{n!} \int_0^\infty s^n e^{-sx} e^{tL(-s)} ds = \frac{1}{n!} \int_0^\infty s^n g(x,t;-s)ds.$

And from $g_+^-(x,t;y) = p_t(y-x) = \frac{1}{2\pi} \int e^{is(x-y)} e^{tL(is)} ds$

we have

(4) $p_{-(n+1)} = \frac{1}{n!} \frac{\partial^n}{\partial y^n}\Big|_{y=0} g_+^- = \frac{(-1)^n}{n!} D^n p_t(-x) = \frac{(-1)^n}{n! 2\pi} \int (is)^n e^{isx+tL(is)} ds$

$= \frac{(-1)^n}{n! 2\pi} \int (is)^n g(x,t;is)ds.$

Examples:

(1) If we let $t = 0$, then the above h's, p's, g's for $t \neq 0$ are obtained by applying e^{tL} to the initial functions. We have:

$$p_0(x) = \delta(x). \quad \Omega_0 = \chi(x).$$

$C(0) = x(0) = x.\quad g = e^{ax}.\quad g_+ = e^{ax}\chi(x).\quad g^- = \frac{1}{x-a}.\quad g_+^- = \delta(a-x).$

$h_n = x^n.\quad h_{-(n+1)} = x^{-(n+1)}.\quad p_n = x^n\chi(x).\quad p_{-(n+1)} = \frac{(-1)^n}{n!}\,\delta^{(n)}(x).$

(2) These L's generate some of the basic stochastic processes:

$L = \frac{z^2}{2}$ Brownian motion $L = \log\cosh z$ symmetric Bernoulli process

$L = e^z - 1$ Poisson process $L = -\log(1-z)$ exponential process

In Chapter IV we will see a "natural" way these arise.

(3) A new process, with p_t a signed measure, that appears at this point, is the Airy process with $L = z^3/3$. Here we remark only that

$p_t(x) = \frac{Ai(xt^{-1/3})}{t^{1/3}}$ form a convolution semigroup with generator $L = z^3/3$.

Ai is the Airy function of the 1st kind. $\mu_{3k}(t) = \frac{3k!}{3^k k!}t^k$, other moments are zero. Thus,

$$h_n = \Sigma\binom{n}{3k}x^{n-3k}t^k\,\frac{3k!}{3^k k!}.$$

INVERSION FORMULA

A very simple observation allows us to map between functions of C and x.

Inversion Principle: Let $F(x,t) = f(C)1$. Then $f(x) = F(\bar{C},t)1$.

Proof: $F(x,t) = e^{tL}f(x).$ $F(\bar{C},t)1 = e^{-tL}F(x,t)e^{tL}1$
$= e^{-tL}e^{tL}f(x) = f(x).$

For example, $x^n = \Sigma\binom{n}{k}\bar{C}^{n-k}1\mu_k(t) = \Sigma\binom{n}{k}\bar{h}_{n-k}(x,t)\mu_k(t).$

Another way of viewing this is:

$$F(x,t) = f[x + tL'(D)]1.\ \text{Thus,}$$
$$F(\bar{C},t)1 = f[(x-tL') + tL']1 = f(x)1 = f(x).$$

In analogy with complex variables, we could define a function of two variables $F(x,t)$ to be analytic (relative to L) if $\frac{\partial}{\partial t}F(\bar{C},t)1 = 0$. Then setting $t = 0$ yields $F(\bar{C},t)1 = F(x,0)$. And $F(\bar{C},t)1 = e^{-tL}F(x,t) = F(x,0) = f(x)$ implies $F(x,t) = e^{tL}f(x)$. That is, "analytic" is the same as "coharmonic."

EXPANSIONS IN TERMS OF h's

In analogy with Taylor series we have the

Expansion Theorem: $$f(x) = \sum_0^\infty \frac{\bar{h}_n(x,t)}{n!} \langle f^{(n)}[w(t)]\rangle$$

or

$$f(x) = \sum_0^\infty \frac{\bar{h}_n(x,t)}{n!} \langle D^n f\rangle.$$

Remark: Recall from Proposition 7 that $\langle F[w(t)]\rangle = e^{tL}F(0) = \int F(y)p_t(y)$.

Proof: $$e^{yD-tL(D)}f(x) = \Sigma \frac{\bar{h}_n(y,t)}{n!} D^n f(x)$$

And $e^{yD-tL(D)}f(x) = e^{-tL}f(x+y)$. Apply e^{tL} to get

$$f(x+y) = \Sigma \frac{\bar{h}_n(y,t)}{n!} e^{tL}D^n f(x), \text{ and set } x = 0.$$

REPRODUCING KERNEL

We next consider an alternative way of expressing a function f; this time by means of a reproducing kernel. Essentially we are looking for an eigenfunction expansion, so in this case we consider the fundamental solution of the equation $\frac{\partial u}{\partial s} + \bar{A}u = 0$.

Definition: $K_t(x,y;\lambda) = \lambda^{\bar{A}}\delta(x-y)$. $(0 < \lambda < 1)$

Remarks: (1) Thus, $K_t(x,y;e^{-s})$ satisfies $\frac{\partial u}{\partial s} + \bar{A}u = 0$, $u(0) = \delta(x-y)$.

(2) The limits $s \to 0$ and $\lambda \to 1$ correspond.

(3) The Green's kernel for $\bar{A}u = f$ is thus $\int_0^\infty K_t(x,y;e^{-s})ds$.

Proposition 14: (1) $$f(x) = \lim_{\lambda\to 1} \int K_t(x,y;\lambda)f(y)dy$$

(2) $$K_t(x,y;\lambda) = \sum_0^\infty \lambda^n \bar{h}_n(x,t)(-1)^n p_{-(n+1)}(-y,t).$$

Remark: (2) shows that K_t is the kernel for expansions in eigenfunctions of $\bar{A}$.

Proof: (1) $$\int K_t(x,y;\lambda)f(y)dy = \lambda^{\bar{A}}f(x) \underset{\lambda\to 1}{\to} f(x).$$

(2) $$\begin{aligned}\lambda^{\bar{A}}\delta(x-y) &= e^{-tL}\lambda^{xD}e^{tL}\delta(x-y)\\ &= e^{-tL}p_t(y-\lambda x) \text{ [Props. 4,7]}\\ &= e^{-tL}\bar{g}_+(-y,t;-\lambda x)\\ &= e^{-tL}\Sigma(-\lambda)^n x^n p_{-(n+1)}(-y,t)\end{aligned}$$

$$= \Sigma(-\lambda)^n \bar{h}_n(x,t)p_{-(n+1)}(-y,t).$$

THE FLOW FOR A

We use the Hamiltonian method to compute another expression for the flow generated by $A = CD$. We have:

$$H = A = CD = [x + tL'(z)](z) = xz + tzL'(z).$$

Remark: Now we fix t and use s as a time variable.

$$\dot{x}(s) = x(s) + tL'[z(s)] + tz(s)L''[z(s)]$$
$$\dot{z}(s) = -z(s).$$

So

$$z(s) = ze^{-s} \quad \text{and} \quad e^{-s}x(s) = x + tL'(z) - te^{-s}L'(ze^{-s}).$$

As above, put $\lambda = e^{-s}$, $x(-\log\lambda) = x(\lambda) = \lambda^{-A}x\lambda^{A}$:

$$\lambda x(\lambda) = x + tL'(z) - \lambda tL'(z\lambda) = C - \lambda tL'(z\lambda)$$

Proposition 15: $\lambda^{A}f(x) = \lambda^{xD}e^{t[L(\lambda D)-L(D)]}f(x)$

Proof: $\lambda^{A} = e^{tL}\lambda^{xD}e^{-tL}$

Using Proposition 4 and EL, observe that

$$e^{aD}\lambda^{xD}f(x) = f(x\lambda + a\lambda) = \lambda^{xD}e^{a\lambda D}f(x)$$

And hence,

$$g(D)\lambda^{xD} = \lambda^{xD}g(\lambda D).$$

Apply this with $g(D) = e^{tL(D)}$.

Proposition 16: $p_t(x)$ is an invariant measure for the flow generated by $\bar{A}$.

Proof:

$$\begin{aligned}\int \lambda^{\bar{A}}f(x)p_t(x) &= e^{tL}\lambda^{\bar{A}}f(0)\\ &= e^{tL}e^{-tL}\lambda^{xD}e^{tL}f(0)\\ &= \lambda^{xD}e^{tL}f(0)\\ &= \int f(\lambda\cdot 0 + y)p_t(y) = \int f(y)p_t(y).\end{aligned}$$

APPLICATION TO THE INTEGRAL OF A PROCESS

We can use the C-operator to deduce

Proposition 17: Let $W(t)$ be a process generated by $L'(z)$. Then

$$\langle \exp[\tfrac{1}{t}\int_0^{at} W(s)ds]\rangle = e^{tL(a)}$$

Proof: The Feynman-Kac formula gives the solution to

$$\frac{\partial u}{\partial a} = xu + tL'(D)u, \quad u(0) = 1$$

as $u = \langle e^{ax}\exp[\int_0^a W(st)ds]\rangle$

and by exponentiating C,

$$u = e^{aC}1 = e^{ax+tL(a)}.$$

By scaling $t \to \frac{1}{b}$, $a \to ab$ we have

Corollary: $\langle \exp[b\int_0^a W(s)ds]\rangle = e^{\frac{1}{b}L(ab)}$.

SOME ASPECTS OF PHYSICAL SIGNIFICANCE

Definition: For L that is invertible as a function of z, define the characteristic operators $\ell = L'^{\circ}L^{(-1)}$, $m = \ell^{(-1)} = L^{\circ}L'^{(-1)}$. The circle denotes composition of functions here.

Time-Space Duality

We can express the flow generated by L in terms of $T = \frac{d}{dt}$. Probabilistically this is essentially fixing a location x and stopping the process the first time it hits x. Thus we have a dual process t(x).

Proposition 18: (1) $e^{tL}g(x) = g(C)1 = g[x + t\ell(T)]1$.

(2) $\langle g[w(t)]\rangle = g[t\ell(T)]1$.

Proof: By EL, $x + t\ell(T) \cdot e^{ax+tL(a)} = x + tL'(a) \cdot e^{ax+tL(a)}$

$$= \partial_a e^{ax+tL(a)}$$

$$= Ce^{aC}1.$$

Thus,

$$g[x + t\ell(T)]e^{ax+tL(a)} = g(\partial_a)e^{ax+tL(a)} = g(C)e^{aC}1.$$

Set a=0 for (1). (2) follows by setting x=0.

The Classical Path

Given L(z) we consider it as the Hamiltonian of a quantum system. We have the equation

$$H\left(\frac{\partial R}{\partial \dot{x}}\right) = \frac{\partial R}{\partial \dot{x}}\dot{x} - R(\dot{x})$$

where R is the associated Lagrangian depending only on $\dot{x}$. It is easy to deduce $H'[R'(\alpha)] = \alpha$, α denoting the argument $\dot{x}$.

Assuming H' is invertible, we have

$$z = R'(\alpha) = H'^{(-1)}(\alpha) = h(\alpha).$$

For $H = L(z)$, m denoting the characteristic operator above,

$$H = m(\dot{x}).$$

Let $q(x)$ denote the potential energy function.

Then $R(x,\dot{x}) = \int^{\dot{x}} h(\alpha)d\alpha - q(x)$.

The new Hamiltonian is $H = m(\dot{x}) + q(x)$.

The classical path is determined by energy = H = constant:

$$H = m(\dot{x}) + q(x) \Rightarrow m(\dot{x}) = H - q(x),$$

and the equation of motion is

$$\dot{x}(t) = \ell[H - q(x(t))].$$

Defining
$$\tau(\alpha) = \int^{\alpha} \frac{dy}{\ell[H-q(y)]},$$

we can integrate the equation of motion, obtaining

$$\tau[x(t)] = \tau(x) + t.$$

When $q=0$, we have

$$\frac{x(t)}{\ell(H)} = \frac{x}{\ell(H)} + t,$$

which is constant velocity motion with speed $\ell(H)$.

Remark: Following Feynman and Kac, we may take the probabilistic "limit" from quantum mechanics, setting $\frac{i}{\hbar} = -1$. Then we may presume that on path space the measure induced by $L(z) + q(x)$ has a "density" of the form

$$\exp\left(-\int_{t_0}^{t} R[\dot{x}(s)]ds + \int_{t_0}^{t} q[x(s)]ds\right)$$

where as above, $R' = L'^{(-1)}$.

EXAMPLES OF MOMENT THEORY

We will consider some natural L's in the context of their moment theory.

1. $L = -\log(1-z)$. $\quad p_t(x) = \dfrac{x^{t-1}e^{-x}\chi(x)}{\Gamma(t)}$. $\quad e^{tL} = (1-z)^{-t}$. $\quad \mu_n(t) = t_{(n)}$.

$$\Omega = \int_{-x}^{\infty} y^{t-1}e^{-y}\chi(y)dy/\Gamma(t) = \chi(x) + [1-\chi(x)]\int_{|x|}^{\infty} y^{t-1}e^{-y}dy/\Gamma(t).$$

$C = x+t(1-z)^{-1}$. $\quad g(x,t;a) = e^{ax}(1-a)^{-t}$. $\quad g_+^-(x,t;a) = p_t(a-x)$.

$g^-(x,t;a) = \int_0^{\infty} e^{ay}e^{-yx}(1+y)^{-t}dy$. Expand e^{ay}, $(1+y)^{-t}$ to get

$$g^-(x,t;a) = \Sigma \frac{a^n}{n!}\int_0^{\infty}\Sigma \frac{(-t)^{(k)}}{k!} y^{k+n}e^{-yx}dy = \sum_n a^n \sum_k \binom{n+k}{k}(-1)^k t_{(k)} x^{-k-n-1}.$$

$$g_+(x,t;a) = \int_0^\infty e^{ay}(y-x)^{t-1}e^{x-y}\chi(y-x)dy/\Gamma(t) = \chi(x)g(x,t;a)$$

$$+ [1-\chi(x)]e^{ax}\int_{|x|}^\infty e^{ay}p_t(y)dy,$$

which reflects the fact that the mass of p_t is concentrated on $(0,\infty)$.
We have then,

$$h_n(x,t) = \Sigma\binom{n}{k}x^{n-k}t_{(k)}. \quad h_{-(n+1)}(x,t) = \Sigma\binom{n+k}{k}(-1)^k t_{(k)}x^{-k-n-1}.$$

$$p_n(x,t) = \chi(x)h_n(x,t) + [1-\chi(x)]\int_{|x|}^\infty (x+y)^n p_t(y) + \text{(terms supported at 0)}.$$

$$p_{-(n+1)} = \frac{(-1)^n}{n!}\left(\frac{\sin 2\pi t}{2\pi}\right)e^x\Sigma\binom{n}{k}\delta^{(n-k-1)}_{(x)}\Sigma\binom{k}{\ell}x^{t-1-\ell}(-1)^\ell\Gamma(\ell-t+1)$$

where we have used

$$D^k\chi(-x) = -\delta^{(k-1)}(x); \quad \frac{(t-1)^{(\ell)}}{\Gamma(t)} = \frac{(-1)^\ell(1-t)_{(\ell)}}{\Gamma(t)} = \frac{(-1)^\ell\Gamma(\ell-t+1)}{\Gamma(t)\Gamma(1-t)}$$

$$= \frac{(-1)^\ell \sin\pi t\,\Gamma(\ell-t+1)}{\pi}$$

$(-1)^t = \cos\pi t$. Note that for $x\neq 0$, only the term $k=n$ is non-zero.
The characteristic operator $\ell(T) = L'^{\circ}L^{-1} = e^T$. So $\mu_n(t) = (te^T)^n 1$.
The Lagrangian $R(\alpha) = \int L'^{-1} = \int 1-\alpha^{-1} = \alpha-\log\alpha-1$.

2. $L = e^z-1$. $\quad p_t(x) = e^{-t}\Sigma\frac{t^k}{k!}\delta(x-k)$. $\quad e^{tL} = e^{t(e^z-1)}$.

$$\mu_n(t) = \partial_z^n\big|_0 e^{tL} = \sum_{k\leq n}\frac{t^k}{k!}\sum_{\substack{r_j>0\\ \Sigma r_j=n}}\frac{n!}{r_1!r_2!\dots r_k!};$$

$\mu_n(1)$ are "Bell numbers."
The coefficient of $\frac{t^k}{k!}$ is the number of ways of grouping n objects into k groups, having at least one object in every group.

$$\Omega = \int_{-x}^\infty p_t(y) = \chi(x) + [1-\chi(x)]e^{-t}\sum_{k\geq|x|}\frac{t^k}{k!}.$$

$$C = x + te^z.$$

We thus have the recursion $h_{n+1}(x,t) = xh_n(x,t) + th_n(x+1,t)$.

$$g(x,t;a) = e^{ax}e^{t(e^a-1)}. \quad g_+^- = p_t(a-x).$$

$$g^- = \int_0^\infty e^{ay}e^{-yx}\exp[t(e^{-y}-1)]dy = \Sigma\frac{a^n}{n!}\int_0^\infty e^{-yx}\,\Sigma\frac{y^{k+n}}{k!}(-1)^k\mu_k(t)$$

$$= \Sigma\, a^n\,\Sigma\binom{n+k}{k}(-1)^k\mu_k(t)x^{-k-n-1}.$$

$$g_+(x,t;a) = \int_0^\infty e^{ay}e^{-t}\Sigma \frac{t^k}{k!}\,\delta(y-x-k)dy = \chi(x)g(x,t;a) + [1-\chi(x)]e^{ax-t}\sum_{k\geq|x|}\frac{t^k}{k!}e^{ak}.$$

We thus have,

$$h_n(x,t) = \Sigma\binom{n}{k}x^{n-k}\mu_k(t). \quad h_{-(n+1)}(x,t) = \Sigma\binom{n+k}{k}(-1)^k\mu_k(t)x^{-k-n-1}.$$

$$p_n(x,t) = \chi(x)h_n(x,t) + [1-\chi(x)]e^{-t}\sum_{k\geq|x|}\frac{t^k}{k!}(k+x)^n + (\text{terms supported at } 0)$$

$$p_{-(n+1)}(x,t) = \frac{(-1)^n}{n!}e^{-t}\Sigma\frac{t^k}{k!}\delta^{(n)}(x+k).$$

The characteristic operator $\ell(T) = L^{\wedge}L^{-1} = 1+T$. We have $\mu_n(t) = [t(1+T)]^n 1$.

The Lagrangian $R(\alpha) = \int L'^{-1} = \int \log\alpha = \alpha(\log\alpha-1)+1$.

3. $L = z^2/2$. $p_t(x) = \frac{1}{\sqrt{2\pi t}}e^{-x^2/2t}$. $e^{tL} = e^{z^2t/2}$. $\mu_n(t) = 0$ for $n = 2k+1$,

$$\mu_{2k}(t) = \frac{(2k)!}{2^k k!}t^k.$$

$\Omega = \Phi(-\frac{x}{\sqrt{t}})$, where $\Phi(x) = \int_x^\infty p_1(y)dy$ is the "complementary error function."

$C = x + tz$ giving a recursion $h_{n+1} = h_n + nth_{n-1}$.

$$g = e^{ax+a^2t/2}. \quad g^- = \int_0^\infty e^{ay-yx}e^{y^2t/2}dy, \text{ diverges for } t > 0.$$

$$\bar{g}^- = \sqrt{\frac{2\pi}{t}}\exp\left(\frac{(x-a)^2}{2t}\right)\Phi\left(\frac{x-a}{\sqrt{t}}\right). \text{ Also } g^- \text{ expands as in the above examples.}$$

$$g^+ = \int_0^\infty e^{ay}p_t(y-x)dy = e^{ax+a^2t/2}\Phi\left(-\frac{x+at}{\sqrt{t}}\right).$$

$$g_+^- = \frac{e^{-(a-x)^2/2t}}{\sqrt{2\pi t}} = p_t(x)\cdot\exp\left[\frac{ax}{t}-\left(\frac{a}{t}\right)^2\frac{t}{2}\right] = p_t(x)\cdot g(x,-t;\tfrac{a}{t}).$$

We thus have

$$h_n(x,t) = \Sigma\binom{n}{2k}t^k\frac{(2k)!}{2^k k!}x^{n-2k} \quad h_{-(n+1)} = \Sigma\binom{n+2k}{2k}\frac{(2k)!}{2^k k!}t^k x^{-2k-n-1}.$$

$$p_n(x,t) = \Sigma\binom{n}{k}(-1)^k t^{k/2}h_{n-k}(x,t)\Phi^{(k)}\left(-\frac{x}{\sqrt{t}}\right)$$

$$p_{-(n+1)}(x,t) = \frac{t^{-n}}{n!}h_n(x,-t)p_t(x).$$

The characteristic operator is $\ell(T) = \sqrt{2T}$. We have $\mu_n(t) = (t\sqrt{2T})^n 1$. This "explains" why the odd moments are zero.

The Lagrangian $R(\alpha) = \int L'^{-1} = \int\alpha = \alpha^2/2$.

4. $L = \log\cosh z.$ $\quad p_t(x) = 2^{-t}\Sigma\binom{t}{k}\delta(x-t+2k).$ $\quad e^{tL} = (\cosh z)^t.$

$$\Omega = \int_{-x}^{\infty} p_t(y) = 2^{-t} \sum_{k \leq \frac{t+x}{2}} \binom{t}{k}. \qquad C = x + t\tanh z.$$

$$g(x,t;a) = e^{ax}\cosh^t a. \qquad g^- = \int_0^{\infty} e^{ay}e^{-yx}\cosh^t y\,dy.$$

$$g_+(x,t;a) = \int_0^{\infty} e^{ay}p_t(y-x)dy = e^{ax}\int_{-x}^{\infty} e^{ay}p_t(y)dy = e^{ax}2^{-t}\sum_{2k \leq t+x}\binom{t}{k}e^{ak}.$$

We thus have

$$h_n(x,t) = \Sigma\binom{n}{k}x^{n-k}\mu_k(t) \qquad h_{-(n+1)}(x,t) = \Sigma\binom{n+k}{k}(-1)^k\mu_k(t)x^{-k-n-1}$$

$$p_n(x,t) = 2^{-t}\sum_{2k\leq t+x}\binom{t}{k}(x+k)^n$$

$$p_{-(n+1)}(x,t) = \frac{(-1)^n 2^{-t}}{n!}\Sigma\binom{t}{k}\delta^{(n)}(x+t-2k)$$

The characteristic operator $\ell(T) = \sqrt{1-e^{-2T}}.$

So we have

$$\mu_n(t) = (t\sqrt{1-e^{-2T}})^n 1. \quad \mu_{2k+1} = 0. \quad \mu_{2k+2}(t) = (t\sqrt{1-e^{-2T}})^2\mu_{2k}(t)$$

$$= (t^2-\tau^2)\mu_{2k}(t) \quad \text{where} \quad \tau = te^{-T}.$$

We derive this as follows:

$$t\sqrt{1-e^{-2T}}\; t\sqrt{1-e^{-2T}} = t^2(1-e^{-2T}) + t\sqrt{1-e^{-2T}}\frac{\partial}{\partial T}\sqrt{1-e^{-2T}}$$

$$= t^2(1-e^{-2T}) + te^{-2T}$$

Observe now that $\tau^2 = te^{-T}te^{-T} = t^2e^{-2T} - te^{-2T}.$

The Lagrangian is $\int \frac{1}{2}\log\frac{1+\alpha}{1-\alpha} = \frac{1}{2}\log(1-\alpha^2) + \frac{\alpha}{2}\log\frac{1+\alpha}{1-\alpha}.$

<u>Remarks</u>: From the above examples, we see that, in general,

(1) $h_{-(n+1)}(x,t) = x^{-n-1}\,\Sigma\binom{n+k}{k}(-1)^k\mu_k(t)x^{-k}.$

(2) If $p_t(x)$ is concentrated on $(0,\infty)$, then for $x > 0, g_+ = g.$

CHAPTER IV. ORTHOGONAL THEORY

In the case $L = \frac{z^2}{2}$, the expansion

$$e^{ax-a^2t/2} = \Sigma \frac{a^n}{n!} H_n(x,t)$$

in Hermite polynomials is not only a moment expansion as in Chapter III, but is an orthogonal expansion as well. We ask: when does the basic harmonic exponential $e^{ax-tL(a)}$ have an expansion in a sequence $J_n(x,t)$ of orthogonal polynomials [relative to $p_t(x)$]? We also look for J_n's generated by Heisenberg duals analogous to C and D. We normalize L so that $L(0) = L'(0) = 0$, $L''(0) = 1$. Then we have the fundamental

Theorem 2 The exponential $e^{a\overline{C}_1}$ has an orthogonal expansion with D mapped into a translation-invariant operator $V(D)$ if and only if L has the form

$$-\frac{\alpha}{\beta}z - \frac{2}{\beta}\log[pe^{Qz/2} + (1-p)e^{-Qz/2}]$$

or a limiting case ($Q=0$; $\beta=0$; $\alpha=\beta=0$)

where α and β are given numbers. $Q = \sqrt{\alpha^2-2\beta}$. $p = \frac{1}{2} - \frac{\alpha}{2Q}$. Furthermore, $V(D) = L'(D)$.

Proof:

Step 1. Set $J_0=1$. We want $V(D)J_n = nJ_{n-1}$. If we have

$$e^{ax-tL(a)} = \sum_0^\infty \frac{b_n(a)}{n!} J_n$$

then

$$V(D)e^{ax-tL(a)} = V(a)e^{ax-tL(a)} = \Sigma \frac{b_n}{n!} nJ_{n-1} = \Sigma \frac{b_{n+1}(a)}{n!} J_n.$$

Thus

$$V(a)\cdot b_n(a) = b_{n+1}(a)$$

or

$$b_n(a) = V(a)^n.$$

Step 2. In order for the J_n's to be orthogonal relative to p_t:

$$\int e^{ax-tL(a)}e^{bx-tL(b)}p_t(x) = \Sigma \frac{[V(a)V(b)]^n}{n!n!} \int J_n^2 p_t = \Sigma \frac{[V(a)V(b)]^n}{n!n!} j_n(t).$$

By definition of L, we thus have

$$e^{t[L(a+b)-L(a)-L(b)]} = F[V(a)V(b)]$$

where

$$F(y) = \Sigma \frac{y^n}{n!} \frac{j_n(t)}{n!}, \qquad j_n(t) = \langle J_n^2 \rangle.$$

This is true if

$$L(a+b) - L(a) - L(b) = \varphi(V_a V_b), \qquad \varphi = \frac{1}{t}\log F.$$

Step 3. Expand to 2nd-order Taylor series around $a = 0$:

$$L(a+b) - L(a) - L(b) = aL'(b) + \frac{a^2}{2}[L''(b)-1] + \ldots$$

$$\varphi(V_a V_b) = aV_b\varphi'(0) + \frac{a^2}{2}[V_b^2\varphi''(0)+V''(0)V_b\varphi'(0)] + \ldots$$

From Step 1 we see, setting $a=0$, that $V(0)=0$. We also see that V can be defined up to a multiplicative constant, so we normalize $V'(0)=1$; this has been used above. Furthermore, φ is important only in that it is a function of the product $V_a V_b$; $\varphi(0)=0$, so we can choose $\varphi'(0)=1$.

Step 4. Then comparing coefficients of a and a^2 we have

$$V(b) = L'(b), \quad L''(b) = 1 + \alpha L'(b) + \frac{\beta}{2}L'(b)^2$$

where $\alpha = V''(0) = L'''(0)$, $\beta = 2\varphi''(0)$.
This yields the theorem.

Remarks:

(1) Differentiating $L'' = 1 + \alpha L' + \frac{\beta}{2}L'^2$ and integrating back gives the equation $L'' = e^{\alpha z+\beta L}$. Thus, L is a convex function for real α,β,z.

(2) The characteristic polynomial for L is defined by $\pi(x) = \beta x^2 + 2\alpha x + 2$. We denote the roots of $\pi(x)$ by $r = \frac{-\alpha+Q}{\beta}$, $s = \frac{-\alpha-Q}{\beta}$. We also use the notations:

$$q = \frac{1}{Q}, \quad a = \alpha q, \quad p = \frac{q\beta r}{2} = \frac{1-a}{2}, \quad \bar{p} = -\frac{q\beta s}{2} = \frac{1+a}{2}.$$

(3) The condition $L'(0) = 0$ means $\langle w(t)\rangle = 0$, i.e. $w(t)$ is centered. It also is the same as $V(D)1 = 0$, which we later want for constructing the canonical calculus where V replaces D.

(4) Since $V'(0) = 1$, V is (locally) invertible near 0. Denote the inverse function by U. Then

$$e^{ax-tL(a)} = \Sigma \frac{V^n(a)}{n!} J_n(x,t)$$

becomes, substituting $a = U(v)$,

$$G(x,t;v) = e^{xU(v)-tM(v)} = \Sigma \frac{v^n}{n!} J_n(x,t), \quad \text{the generating}$$

function for the J_n's. Note that $M(v) = L[U(v)] = m(v)$ the 2nd characteristic operator for L.

CANONICAL FORMS

We will now see that we can focus on five standard processes. $w(t)$ will denote the process directly generated by L.

1. $\alpha,\beta,Q \neq 0$. $L = -\frac{\alpha}{\beta}z - \frac{2}{\beta}\log[pe^{Qz/2}+(1-p)e^{-Qz/2}]$. We scale $T = -\frac{2t}{\beta}$, $X = 2qx$, and define $x(t)$ by

$$\frac{1}{t}\log\langle e^{zx(t)}\rangle = \log(pe^{z}+\bar{p}e^{-z}) \quad \text{(see notation above).}$$

Then $w(t) = \frac{1}{2q}\,x(-\frac{2t}{\beta})-\frac{\alpha t}{\beta}$. And it is clear that $x(t)$ is distributed as the t-fold convolution of a Bernoulli measure with mass p at +1, $\bar{p}$ at -1. In fact,

$$\int e^{2x}p_t(x) = (pe^{z}+\bar{p}e^{-z})^t = \Sigma\frac{t^{(k)}}{k!}\,\bar{p}^k p^{t-k} e^{z(t-2k)}$$
$$= \int e^{zx}\Sigma\binom{t}{k}\bar{p}^k p^{t-k}\delta[x-(t-2k)].$$

We call $x(t)$ the Bernoulli process.

2. $\alpha = 0,\beta,Q \neq 0$. Set $b = -\beta$. Scale $T = \frac{2t}{b}$. $X = \sqrt{2/b}\,x$.
Define $d(t)$, the symmetric Bernoulli process, by

$$\frac{1}{t}\log\langle e^{zd(t)}\rangle = \log\cosh z.$$

Then $w(t) = \sqrt{\frac{b}{2}}\,d(\frac{2t}{b})$. Here $p = \bar{p} = \frac{1}{2}$. $p_t(x) = 2^{-t}\Sigma\binom{t}{k}\delta[x-(t-2k)]$.

3. $\alpha,\beta \neq 0$, $Q = 0$. $\pi(x)$ is a perfect square; $r = s = -\frac{\alpha}{\beta}$.
L reduces to $rz-r^2\log(1+\frac{z}{r})$. We scale $T = r^2t$, $X = -rx$.
Define $e(t)$, the exponential process, by

$$\frac{1}{t}\log\langle e^{ze(t)}\rangle = -\log(1-z).$$

Then $w(t) = -\frac{1}{r}[e(r^2t)-r^2t]$ and $p_t(x) = \frac{x^{t-1}e^{-x}}{\Gamma(t)}\,\chi(x)$ are the gamma densities.

4. $\beta = 0$. $Q = \alpha$. $\pi(x)$ is linear. L is $\frac{1}{\alpha^2}(e^{\alpha z}-1-\alpha z)$.

We scale $T = \frac{t}{\alpha^2}$, $X = \frac{1}{\alpha}x$. Then define the Poisson process $N(t)$, as usual, by

$$\frac{1}{t}\log\langle e^{zN(t)}\rangle = e^z-1.$$

Then $w(t) = \alpha[N(\frac{t}{\alpha^2})-\frac{t}{\alpha^2}]$. And $p_t(x) = e^{-t}\sum_0^\infty\frac{t^k}{k!}\,\delta(x-k)$.

5. $\alpha = \beta = 0$. L reduces to $\frac{z^2}{2}$. Brownian motion $b(t)$ satisfies

$$\frac{1}{t}\log\langle e^{zb(t)}\rangle = \frac{z^2}{2}.$$

We have $w(t) = b(t)$; and $p_t(x) = \frac{e^{-x^2/2t}}{\sqrt{2\pi t}}$.

In Chapter V we will study these in more detail. Presently we proceed with the theory.

We define as in Proposition 2, $W(D) = \frac{1}{V'(D)}$ and the canonical dual $\xi = xW$.

Proposition 19: Set $\bar{\xi}(t) = e^{-tL}\xi e^{tL} = \bar{C}W$. Then

(1) $J_n(x,t) = \bar{\xi}(t)^n 1$

(2) $\bar{C}WJ_n(x,t) = J_{n+1}(x,t)$

(3) $V(D)J_n(x,t) = nJ_{n-1}(x,t)$

(4) $\bar{\xi}(t)VJ_n = \bar{C}WVJ_n = nJ_n$

(5) $G(x,t;v) = e^{v\bar{C}W}1$

(6) J_n are the moment polynomials in ξ relative to the generator $M(V)$ (i.e. $e^{-tM(V)}\xi e^{tM(V)} = e^{-tL}xWe^{tL}$).

(7) $J_n(x,t) = \Sigma\binom{n}{k}\xi^{n-k}1\mu_k(-t)$, where μ_k are the moments corresponding to the generator $M(V)$.

Proof: By Proposition 3,

$$e^{v\xi}1 = e^{xU[v+V(0)]} = e^{xU(v)}.$$

So,

$$e^{v\bar{\xi}(t)}1 = e^{-tL}e^{xU(v)} = e^{xU(v)-tM(v)} = G(x,t;v).$$

We can also check directly from $G = e^{xU-tM}$ that $\frac{\partial G}{\partial v} = \bar{C}WG$, $G(0) = 1$. 1 is a vacuum function since $V(0) = 0$, $W(0) = 1$; so $WV1 = W(0)V(0) = 0$. The rest follows by the theory of Chapter III.

Remarks: (1) See Proposition 26 for more concerning (6).

(2) It is easy to check that in canonical variables $\bar{\xi}(t) = \xi - t\,\frac{2V}{\pi(V)}$.

REPRESENTATIONS OF J_n

As in Chapter III we have

$$J_n(x,t) = \partial_v^n\big|_0 G(x,t;v)$$

$$= \frac{n!}{2\pi i}\int_0 \frac{e^{xU(v)-tM(v)}}{v^{n+1}}\,dv.$$

Since $V(0) = 0$, $V'(0) = 1$, we can change variables $v = V(z)$ and still be on a 0 contour. We get

$$J_n(x,t) = \frac{n!}{2\pi i}\int_0 \frac{e^{xz-tL(z)}}{V^{n+1}(z)}\,\frac{dz}{W(z)}$$

Now recall that $L''(z) = e^{\alpha z+\beta L} = \frac{1}{W(z)}$. So

$$J_n(x,t) = \frac{n!}{2\pi i}\int_0 \frac{e^{z(x+\alpha)-(t-\beta)L(z)}}{V(z)^{n+1}}\,dz$$

$$= \partial_z^n\big|_0 \left(\frac{z}{V(z)}\right)^{n+1} \overline{g}(x+\alpha, t-\beta; z)$$

which also

$$= e^{\alpha D-(t-\beta)L(D)} \left(\frac{D}{V(D)}\right)^{n+1} x^n \quad \text{by the duality lemma}$$

$$= e^{-tL(D)} \frac{1}{W(D)} \left(\frac{D}{V(D)}\right)^{n+1} x^n$$

and $$= \left(\frac{D}{V(D)}\right)^{n+1} \overline{h}_n(x+\alpha, t-\beta).$$

This thus gives us the mapping between the orthogonal and moment systems for L.

EXPANSIONS

First Expansion Theorem

$$f(x) = \sum_0^\infty \frac{J_n(x,t)}{n!} \langle V^n f[w(t)]\rangle$$

or

$$f(x+a) = \sum_0^\infty \frac{J_n(x,t)}{n!} e^{tL} V^n f(a).$$

This follows exactly as for the moment expansion. Surprisingly we have derived the orthogonal expansion of f by what is essentially a Taylor expansion. We will now find a richer structure here than for moment expansions because we do have orthogonality.

GENERALIZED RODRIGUES FORMULA

We will find that Rodrigues' Formula for the J_n's is actually a special case of a general inversion formula. We define the transform of f, f*, to be the generating function of $f_n = \langle fJ_n\rangle$, the unnormalized Fourier coefficients of f, i.e. $f_n = \int f(y)J_n(y,t)p_t(y)$.

Proposition 20: (1) $f^* = \Sigma \frac{x^n}{n!} f_n$

(2) $f^* = \langle G[w(t),t;x]f[w(t)]\rangle = \int G(y,t;x)f(y)p_t(y)$.

Proof: (1) is the definition. For (2):

$$f^* = \Sigma \frac{x^n}{n!} \int f(y)J_n(y,t)p_t(y) = \int G(y,t;x)f(y)p_t(y).$$

Theorem 3 Generalized Rodrigues' Formula (GRF).

Let f* = transform of f. V* = adjoint of V = V(-D).

Then

$$f(x) = \frac{1}{p_t(x)} f^*(V^*)p_t(x).$$

__Proof:__ $$f^*(V^*)p_t = \Sigma\langle fJ_n\rangle \frac{V^{*n}}{n!} p_t$$

Note that $\langle J_n\rangle = 0$ for $n > 0$ since $\langle e^{ax(t)-tL(a)}\rangle = 1 = \Sigma \frac{V^n(a)}{n!} \langle J_n\rangle$.

Now we check that

$$\int [f^*(V^*)p_t]J_k = \Sigma \frac{f_n}{n!} \int J_k V^{*n} p_t$$

$$= \Sigma \frac{f_n}{n!} \int (V^n J_k)p_t$$

$$= \frac{f_k}{k!} \cdot k! \int p_t = f_k$$

$$\text{since } V^n J_k = \begin{cases} 0 & n > k \\ k^{(n)} J_{k-n} & n \le k \end{cases} \qquad \left(\text{recall } k^{(n)} = \frac{k!}{(k-n)!}\right).$$

__Remark:__ The functions of Proposition 1 are synthesizable in terms of the J_k's: since $\deg(J_k) = k$, we can get all polynomials and so on. Thus in the above proof it is enough to check Fourier coefficients.

The next theorem makes explicit the "Fourier vs. Taylor" duality. We first find Rodrigues' formula for J_n.

__Definition:__ $j_n(t) = \langle J_n^2\rangle = \text{var } J_n = \int J_n^2(y)p_t(y)$.

__Proposition 21:__ (Rodrigues' Formula for J_n)

$$J_n = \frac{j_n}{n!} \frac{1}{p_t(x)} V^{*n} p_t(x).$$

__Proof:__ Apply GRF with $f = J_n$. Then

$$f_k = \langle J_k J_n\rangle = j_n \delta_{kn} \quad \text{and} \quad f^* = \frac{x^n}{n!} j_n.$$

Now we have the

__Duality Theorem__ $\langle fJ_n\rangle = \frac{j_n}{n!}\langle V^n f\rangle.$

__Proof:__

1st proof: $\int fJ_n p_t = \frac{j_n}{n!} \int f V^{*n} p_t = \frac{j_n}{n!} \int V^n f \cdot p_t$ by Prop. 21 and definition of V^*.

2nd Proof: Compare the usual orthogonal expansion

$$f(x) = \Sigma \frac{J_n(x,t)}{j_n} \langle fJ_n \rangle$$

with the one in the first expansion theorem.

Data for J_n's: Recurrence Formula. Value of j_n.

We first calculate

Proposition 22: $J_0 = 1 \qquad J_1 = x \qquad J_2 = x^2-\alpha x-t$

Proof: $J_0 = 1$ by construction. $J_1 = \overline{C}W1 = xW-tVW1 = xW(0) = x.$

$$J_2 = (xW-tVW)x = xWx-tVWx = [x^2W+xW'-txVW-t(V'W+VW')]1$$
$$= x^2W(0)+xW'(0)-tV'(0)W(0).$$

(We used GLM and $g(D)1 = g(0)$ here.)

Recall that $W(0) = L''(0) = 1$ and $W'(0) = -\frac{L'''(0)}{L''(0)^2} = -\alpha$ (see remark 1 following or Step 4 of the proof of the main Theorem 2).

Recurrence Formula $\qquad J_{n+1} = (x-\alpha n)J_n - n(t+\frac{n-1}{2}\beta)J_{n-1}$

Proof: We calculate for $k \leq n$, "x" denoting $w(t)$,

$$\langle (J_{n+1}-xJ_n)J_k\rangle = -\langle xJ_nJ_k\rangle.$$

We have:

(1) zero for $k < n-1$: since $\deg(xJ_k) < n$ it can be expressed in terms of J's with index $< n$ and so is $\perp J_n$.

(2) For $k = n$, $\frac{n!}{j_n}\langle xJ_nJ_k\rangle = \langle V^n xJ_n\rangle = \langle xV^nJ_n\rangle + \langle nV^{n-1}V'J_n\rangle$ by GLM

$$= n!\langle x\rangle + \langle n\cdot n!\cdot(e^{\alpha D+\beta L}x)\rangle$$
$$= 0 + n!n\cdot\langle J_1(x+\alpha,t-\beta)\rangle$$
$$= \alpha n\cdot n!, \text{ using Prop. 22.}$$

(3) For $k = n-1$, $\frac{n!}{j_n}\langle xJ_nJ_{n-1}\rangle = \langle V^n xJ_{n-1}\rangle = \langle nV^{n-1}V'J_{n-1}\rangle$

$$= n!$$

We can also calculate this case:

$$\langle xJ_nJ_{n-1}\rangle = \frac{j_{n-1}}{(n-1)!}\langle V^{n-1}xJ_n\rangle$$
$$= \frac{j_{n-1}}{(n-1)!}\left(\langle xV^{n-1}J_n\rangle + (n-1)\langle V^{n-2}V'J_n\rangle\right)$$

And $\langle xV^{n-1}J_n\rangle = n!\langle x^2\rangle = n!t$

$$\langle V'V^{n-2}J_n\rangle = \frac{n!}{2}\langle V'J_2\rangle = \frac{n!}{2}\langle e^{\alpha D+\beta L}J_2\rangle$$
$$= \frac{n!}{2}\langle J_2(x+\alpha, t-\beta)\rangle$$
$$= \frac{n!}{2}\langle (x+\alpha)x-t+\beta\rangle$$
$$= \frac{n!}{2}\beta.$$

Summarizing the above results:

$$\langle xJ_nJ_n\rangle = \alpha\, n j_n$$

$$\langle xJ_nJ_{n-1}\rangle = j_n = j_{n-1}(nt + \frac{n(n-1)}{2}\beta).$$

The formula now follows.

From $j_o = 1$ and the last equality above:

Corollary: $j_n(t) = n!t(t+\beta/2)\ldots(t+\frac{n-1}{2}\beta)$

or $j_n(t) = n!(\frac{\beta}{2})^n(\frac{2t}{\beta})_{(n)}$

Remarks: (1) The recursion yields, for example, $J_3 = x(x-\alpha)(x-2\alpha)-3tx+2\alpha t-\beta x$. And, as may be expected, J_2 and J_3 determine α and β.

(2) We have now that $\langle G[w(t),t;v]^2\rangle = \Sigma \frac{v^{2n}}{n!n!} j_n(t)$

$$= e^{tL[2U(v)]-2tL[U(v)]}$$
$$= (1-\frac{\beta}{2}v^2)^{-2t/\beta}.$$

REPRODUCING KERNEL AND ASSOCIATED KERNEL

Since the J_n's are an orthogonal system we can immediately write

$$K_t(x,y;\lambda) = \sum_o^\infty \frac{\lambda^n J_n(x,t)J_n(y,t)}{j_n(t)}.$$

We also consider the associated kernel

$$k_t(x,y) = \Sigma \frac{y^n J_n(x,t)}{j_n(t)}.$$

Definition: $\psi_t(x) = \sum_o^\infty \frac{x^n}{j_n(t)}.$

<u>Proposition 23</u>: Properties of ψ, k, K.

(1) $\psi_t(x) = \Gamma(s)X^{(1-s)/2}I_{s-1}(2\sqrt{X})$ where $s = \frac{2t}{\beta}$, $X = \frac{2x}{\beta}$, $I_\nu(x)$ is the usual modified Bessel function.

(2) $\psi_t(xy)$ satisfies $\frac{\beta}{2}xf'' + tf' = yf$.

(3) k_t satisfies $\frac{\beta}{2}\xi V^2 k + (tV-y)k = \beta t\frac{V^3}{\pi(V)}k$

(4) $k_t(x,y) = e^{-tM(V)}\psi_t(y\xi)1$

(5) Denote by η, ξ expressed in the variable y; M(x), M(y) denote M(V) acting on x or y variables respectively. Then

$$K_t(x,y;\lambda) = e^{-tM(x)}e^{-tM(y)}\psi_t(\lambda\xi\eta)1 = e^{-tM(y)}k_t(x,\lambda\eta)1.$$

<u>Proof</u>: (1) is the definition of I_ν; use $j_n(t) = n!(\frac{\beta}{2})^n(\frac{2t}{\beta})_{(n)}$.

(2) can be checked directly by definition of ψ.

(3) Apply e^{-tM} to equation (2), using (4):

$$\frac{\beta}{2}\xi V^2\psi_t(y\xi)1 + (tV-y)\psi_t(y\xi)1 = 0$$

so, $\frac{\beta}{2}\xi V^2 e^{-tM}\psi_t(y\xi)1 + (tV-y)e^{-tM}\psi_t(y\xi)1 = -\frac{\beta}{2}[e^{-tM},\xi]V^2\psi_t(y\xi)1.$

Since $[e^{-tM},\xi] = -tM'e^{-tM} = -\frac{2tV}{\pi(V)}e^{-tM}$, we get

$$\frac{\beta}{2}\xi V^2 k + (tV-y)k = \frac{\beta t}{\pi(V)}V^3 k.$$

(4) $e^{-tM}\psi_t(y\xi)1 = \Sigma\frac{y^n J_n}{j_n} = k_t$.

(5) is similar to (4).

We also have formulas for the actions of V and $\overline{C}W$ on k and K.

<u>Proposition 24</u>: (1) $Vk_t = \frac{y}{t}e^{\beta L/2}k_{t+\beta/2}$. $\overline{C}Wk_t = e^{-\beta L/2}(t-\beta/2)\frac{\partial}{\partial y}k_{t-\beta/2}$.

(2) $V_xK_t = \frac{\lambda}{t}e^{\beta/2(L_x+L_y)}\overline{C}W(y,t+\beta/2)K_{t+\beta/2}$

$$\lambda\overline{C}W(x,t)K_t = e^{-\beta/2(L_x+L_y)}(t-\beta/2)V_yK_{t-\beta/2}$$

<u>Remark</u>: The subscripts and arguments denote the variables associated with the indicated operators.

<u>Proof</u>: (1) follows from $VJ_n = nJ_{n-1}$, $\overline{C}WJ_n = J_{n+1}$.

(2) follows similarly.

We also see that, as in Proposition 14, $K_t(x,y;e^{-s})p_t(y)$ is the fundamental solution to $\frac{\partial u}{\partial s} + \overline{C}WVu = 0$. Namely, we have

<u>Proposition 25</u>: (1) $K_t(x,y;\lambda)p_t(y) = \lambda^{\overline{C}WV}\delta(x-y)$.

(2) $\lambda^{\overline{C}WV}f(x) = \langle K_t f\rangle$.

(3) $f(x) = \lim_{\lambda\to 1} \langle K_t f\rangle = \lim_{\lambda\to 1} \int K_t(x,y;\lambda)f(y)p_t(y)$

(4) $\int_0^\infty K_t(x,y;e^{-s})ds\ p_t(y)$ is the Green's kernel for $\overline{C}WVu = f$.

<u>Proof</u>:
$$\lambda^{\overline{C}WV}f(x) = \lambda^{\overline{C}WV}\ \Sigma\ \frac{J_n(x,t)}{j_n}\langle fJ_n\rangle = \Sigma\ \frac{\lambda^n J_n}{j_n} f_n.$$

That is,

$$\int K_t(x,y;\lambda)p_t(y)f(y) = \Sigma \frac{\lambda^n J_n(x,t)}{j_n} f_n = \lambda^{\overline{C}WV}f(x).$$

(1)-(4) follow.

CANONICAL VARIABLES

We make some further remarks regarding the representation in terms of ξ and V. First we check that indeed the action of M(V) on functions of ξ is the same as that of L(D) on functions of xW. We check on exponentials.

<u>Proposition 26</u>: $e^{aM(V)}e^{b\xi}e^{cx} = e^{aL}e^{bxW}e^{cx}$.

<u>Proof</u>: First observe that from

$$e^{aD_o}\ e^{bx} = e^{bx}e^{a(D+b)}$$

we deduce

$$g(D)^o e^{bx} = e^{bx}g(D+b) \quad \text{(This also follows from GLM).}$$

So for the left-hand side we have (using Proposition 3 and EL)

$$\begin{aligned} e^{b\xi}e^{aM(b+V)}e^{cx} &= e^{b\xi}e^{cx}e^{aM[b+V(c)]} \\ &= e^{xU[b+V(c)]}e^{aM[b+V(c)]} \\ &= e^{aL(D)}e^{xU[b+V(c)]} \quad \text{since } M = L^oU \\ &= e^{aL}e^{bxW}e^{cx}. \end{aligned}$$

We can consider now the process $\omega(t)$ with density $p_t(\xi)$ such that

$$\langle e^{a\omega(t)}\rangle = e^{tM(a)} = \int e^{a\xi}p_t(\xi)$$

The basic powers are ξ^n and the generating function for ξ^n

$$e^{a\xi}, \text{ with } e^{a\xi}1 = e^{xU(a)},$$

replaces the exponential e^{ax}. The usual formula for $\bar{h}_n$ and the inversion principle gives us the relations

(1) $J_n(x,t) = \Sigma\binom{n}{k}\xi^{n-k}1\mu_k(-t)$

(2) $\xi^n 1 = J_n(x,0) = \Sigma\binom{n}{k}J_k(x,t)\mu_k(t)$

with $\mu_k(t) = k^{th}$ moment of $p_t(\xi) = \partial_v^k|_0 e^{tM(v)} = J_k(0,-t)$.

We thus have a distribution in the operator variable ξ.

<u>Proposition 27</u>: We can represent

$$p_t(\xi) = \frac{1}{2\pi}\int_U e^{-\xi V(iz)}e^{tL(iz)}\frac{dz}{W(iz)}$$

where U is the contour $-iU(i\mathbb{R})$.

<u>Proof</u>: As in the representation for J_n, substitute $iv = V(iz)$, $z = -iU(iv)$, in

$$p_t(\xi) = \frac{1}{2\pi}\int e^{-i\xi v}e^{tM(iv)}dv. \quad \text{We get}$$

$$p_t(\xi) = \frac{1}{2\pi}\int_U e^{-\xi V(iz)}e^{tL(iz)}\frac{dz}{W(iz)}$$

or

$$= \frac{1}{2\pi}\int_U e^{-i\alpha z-\xi V(iz)+(t-\beta)L(iz)}dz.$$

Corresponding to $p_t(\xi)$ we have the moment-theory generating functions g, g^-, g_+, g_+^-. We define the corresponding functions of x, $G = \bar{g}(\xi)1$, $G^- = g^-1$, $G_+ = g_+1$, $G_+^- = g_+^-1$. Note that we use the conjugate g <u>only</u> in defining G. Our basic g's are

$$g(x,t;z) = e^{zx+tL(z)}, \quad g(\xi,t;v) = e^{v\xi+tM(v)}.$$

We can find expressions for the G's using our operators V and W. We define the contours $U^- = -U(-\infty,0)$, $U = -iU(i\mathbb{R})$.

<u>Proposition 28</u>: Define $\delta_U(x) = \frac{1}{2\pi}\int_U e^{isx}ds$. We have

(1) $G(x,t;v) = \bar{g}(\xi,t;v)1 = e^{v\xi-tM(v)}1 = e^{xU(v)-tM(v)}$.

(2) $G_+^-(x,t;y) = p_t(y-\xi)1 = e^{-yV}W^{-1}\int_U g(x,t;is)ds/2\pi$

$$= e^{tL}e^{-yV}W^{-1}\delta_U(x).$$

(3) $G^-(x,t;y) = e^{-yV}W^{-1}\int_{U^-} g(x,t;-z)dz.$

(4) $G_+(x,t;y) = e^{tL}(V-y)^{-1}W^{-1}\delta_U(x)$.

Proof: (1) has been considered above. For (2),

$$p_t(y-\xi)1 = \frac{1}{2\pi}\int e^{-ys}e^{is\xi}e^{tM(is)}ds1$$

$$= \frac{1}{2\pi}\int e^{-iys}e^{xU(is)+tM(is)}ds$$

$$= e^{tL}e^{-yV}\frac{1}{2\pi}\int_U e^{isx}\frac{ds}{W(is)}, \text{ as in Proposition 27.}$$

For (3), $G^-(x,t;y) = \int_0^\infty e^{ys}e^{xU(-s)+tM(-s)}ds$. Put $U(-s) = -z$.

For (4), $G_+(x,t;y) = \int_0^\infty e^{ys}p_t(s-\xi)1ds$

$$= \int_0^\infty e^{ys}e^{tL}e^{-sV}W^{-1}\delta_U(x)ds, \text{ from (2)}$$

$$= e^{tL}(V-y)^{-1}W^{-1}\delta_U(x).$$

In the following we will study the explicit forms the J's and associated operators and functions take. We will find the representations in canonical variables; and indicate some applications of the theory. In Chapter VI we will study some interrelationships among our five processes in the context of limit theorems.

APPENDIX TO CHAPTER IV

SUMMARY OF NOTATION USED FOR PARAMETERS AND GENERAL FORMS

$w(t)$ corresponds to the generator.

$$-\frac{\alpha}{\beta}z - \frac{2}{\beta}\log(pe^{Qz/2} + \bar{p}e^{-Qz/2})$$

α,β are given. $Q = \sqrt{a^2-2\beta}$. $qQ = 1$.

$$\pi(x) = \beta x^2 + 2\alpha x + 2 \text{ has roots } r = \frac{-\alpha+Q}{\beta},\ s = \frac{-\alpha-Q}{\beta}.$$

$$a = \alpha q. \quad p = \frac{1-a}{2} = \frac{q}{s}. \quad \bar{p} = \frac{1+a}{2} = -\frac{q}{r}.$$

The general form of G is

$$G(x,t;v) = e^{xU(v)-tM(v)} = (1-\frac{v}{r})^{qx-qrt}(1-\frac{v}{s})^{qst-qx}$$

so $$M(v) = \log[(1-\frac{v}{r})^{qr}(1-\frac{v}{s})^{-qs}].$$

Differentiating yields

$$J_n(x,t) = (-1)^n\Sigma\binom{n}{k}r^{k-n}s^{-k}(qx-qrt)^{(n-k)}(qst-qx)^{(k)}$$

where $y^{(k)}$ denotes the factorial power $y(y-1)\ldots(y-k+1) = (-1)^k(-y)_{(k)}$.

Differentiating L yields the general

$$V(z) = \frac{s\partial_Q^+}{s+\partial_Q^+} \quad \text{and} \quad U(v) = q\log\frac{1-\frac{v}{r}}{1-\frac{v}{s}}$$

where $\partial_Q^{\pm}(z) = \frac{e^{\pm Qz}-1}{\pm Q}$ are the basic difference operators.

The canonical variable $\xi = xW = x[1+\frac{1}{s}\partial_Q^+(z)]^2 e^{-Qz}$.

The eigenfunction ("number") operator

$$\overline{C}WV = x\partial_Q^- + (\frac{x}{s}-t)\partial_Q^+\partial_Q^-$$

is a discrete version of the (modified) confluent hypergeometric operator

$$x\frac{\partial}{\partial x} + (\frac{x}{s}-t)\frac{\partial^2}{\partial x^2}.$$

CHAPTER V. THE STANDARD PROCESSES. EXAMPLES.

We begin with a modified version of the general w-process. We will then specialize to the various processes obtained as limits of the general one.

<u>Remarks</u>: (1) The appendix to Chapter IV gives the basic notation and general formulas.

(2) Propositions 27 and 28 are the basis for the expressions for the generating functions. We will derive these for each of the standard processes.

(3) In indicating changes of variables, we use primes to denote the "old" variables. E.g., $f(y)$, $y' = 3y$ becomes $f(3y)$.

THE BERNOULLI PROCESS

1. Define $x(t)$ by $\frac{1}{t}\log\langle e^{zx(t)}\rangle = \log(pe^{z}+\bar{p}e^{-z})$, $p = \frac{1}{2}(1-\alpha q)$, $\bar{p} = 1-p$.

Then $w(t) = \frac{1}{2q}x(-\frac{2t}{\beta})-\frac{\alpha t}{\beta}$, $x(t) = 2qw(-\frac{\beta t}{2})-at$ $(a = \alpha q)$.

$p_t(x) = \Sigma\binom{t}{k}p^{t-k}\bar{p}^{k}\delta(x-t+2k)$.

Given "a" we define the following "asymmetric" functions:

<u>Definition</u>: (1) $\mathrm{snh}\, z = pe^{z}-\bar{p}e^{-z}$. $\mathrm{csh}\, z = pe^{z}+\bar{p}e^{-z}$.

(2) $i\,\mathrm{sn}\, z = \mathrm{snh}(iz)$. $\mathrm{cs}\, z = \mathrm{csh}\, iz$.

(3) $\varepsilon^2 = 4p\bar{p}$, $k^2 = \frac{\bar{p}}{p}$, $K^2 = \frac{p}{\bar{p}}$ are moduli that reduce to 1 in the symmetric case $p = \bar{p} = \frac{1}{2}$.

(4) $T(z) = \varepsilon\frac{\sin z}{\mathrm{cs}\, z}$.

<u>Remarks</u>: (1) We have $\mathrm{sn} z = \sin z + ia\cos z$ $\qquad$ $\mathrm{snh} z = \sinh z - a\cosh z$

$\mathrm{cs} z = \cos z - ia\sin z$ $\qquad$ $\mathrm{csh} z = \cosh z - a\sinh z$

(2) $\mathrm{sn} z \sin z + \mathrm{cs} z \cos z = 1$. $\mathrm{csh} z \cosh z - \mathrm{snh} z \sinh z = 1$.

(3) $L(z) = \log \mathrm{csh} z$ is thus the generator for $x(t)$.

(4) $\varepsilon^2 = 1-a^2$. Put $a = \cos\varphi$, $\varepsilon = \sin\varphi$. Then $p = \sin^2\varphi/2$, $\bar{p} = \cos^2\varphi/2$, $k = \cot\varphi/2$, $K = \tan\varphi/2$.

2. Scaling and Basic Formulas for $x(t)$.

The mapping $v = \frac{\varepsilon}{2q}v'$, $t = -\frac{2t'}{\beta}$, $x = 2qx'-at$ yields the following:

$$G(x,t;v) = (1+kv)^{\frac{t+x}{2}}(1-Kv)^{\frac{t-x}{2}}.$$

$$U(v) = \tfrac{1}{2}\log\frac{1+kv}{1-Kv}, \quad V(z) = \varepsilon\frac{\sinh z}{\mathrm{csh} z} = -iT(iz).$$

$$M(v) = -\tfrac{1}{2}\log(1+kv)(1-Kv) = -\tfrac{1}{2}\log(1+2v\cot\varphi-v^2).$$

$W(z) = \frac{1}{\varepsilon}\mathrm{csh}^2 z$ and $\xi = xW = \frac{1}{\varepsilon}x\ \mathrm{csh}^2 z$ is the canonical dual to V.

$$\bar{C}W = (x-tL')W = x \csc\varphi \operatorname{csh}^2 z - t \csc\varphi \operatorname{snh} z \operatorname{csh} z.$$

Remark: Note that because of the drift in $x(t)$, $V \neq L'$ here.

3. Formulas for J_n, $\xi^n 1$, moments.

From the expression for G we have

$$J_n(x,t) = \Sigma\binom{n}{\ell}(-1)^{n-\ell}K^{n-2\ell}\left(\frac{t+x}{2}\right)^{(\ell)}\left(\frac{t-x}{2}\right)^{(n-\ell)}.$$

$$\xi^n 1 = J_n(x,0) = \Sigma\binom{n}{\ell}K^{n-2\ell}\left(\frac{x}{2}\right)^{(\ell)}\left(\frac{x}{2}\right)_{(n-\ell)}.$$

$$\mu_n(t) = J_n(0,-t) = \Sigma\binom{n}{\ell}(-1)^{\ell}K^{n-2\ell}\left(\frac{t}{2}\right)_{(\ell)}\left(\frac{t}{2}\right)_{(n-\ell)}.$$

Introduce the Gegenbauer polynomials C_n^s defined by

$$(1-2xy+y^2)^{-s} = \sum_0^\infty C_n^s(x)y^n.$$

Then $$e^{tM(v)} = (1+2v\cot\varphi - v^2)^{-t/2} = \sum_0^\infty C_n^{t/2}(i\cot\varphi)\cdot(iv)^n.$$

And so we have

$$\mu_n(t) = n!\, i^n C_n^{t/2}(i\cot\varphi).$$

4. Canonical density. Canonical generating functions.

Proposition 27 gives us

$$p_t(\xi) = \frac{1}{2\pi}\int_U e^{-\xi V(iz)}e^{tL(iz)}\frac{dz}{W(iz)}$$

$$= \frac{1}{2\pi}\int_{-\pi/2}^{\pi/2} e^{-i\xi T(\theta)}\operatorname{cs}^{t-2}\theta\, \mathcal{E}\, d\theta.$$

We can also calculate directly

$$p_t(\xi) = \frac{1}{2\pi}\int e^{-i\xi v}(1+2i(\cot\varphi)v+v^2)^{-t/2}\,dv$$

$$= \frac{1}{2\pi}\mathcal{E}^{t-1}e^{-\xi\cot\varphi}\int e^{-i\xi v\csc\varphi}(v^2+1)^{-t/2}dv \quad \text{(using Cauchy's Theorem)}$$

$$= \frac{\sin^\nu\varphi}{\sqrt{\pi}\,\Gamma(t/2)}\left(\frac{\xi}{2}\right)^\nu e^{-\xi\cot\varphi}K_\nu(\xi\csc\varphi)$$

where $\nu = \frac{t-1}{2}$ and K_ν is MacDonald's (modified Bessel) function.

We have $$g_+^- = p_t(y-\xi) = \frac{1}{2\pi}\int_{-\pi/2}^{\pi/2} e^{-iyT(\theta)}e^{i\xi T(\theta)}\operatorname{cs}^{t-2}\theta\, \mathcal{E}\, d\theta$$

and $$G_+^- = g_+^- 1 = \frac{1}{2\pi}\int_{-\pi/2}^{\pi/2} e^{-iyT(\theta)}\operatorname{cs}^{t-2}\theta\, e^{xU[iT(\theta)]}\mathcal{E}\, d\theta$$

$$= \frac{1}{2\pi}\int_{-\pi/2}^{\pi/2} e^{-iyT(\theta)}e^{ix\theta}\operatorname{cs}^{t-2}\theta\, \mathcal{E}\, d\theta.$$

Next we have

$$g_+(\xi,t;y) = \int_0^\infty e^{ys}p_t(s-\xi)ds = \frac{1}{2\pi}\int_{-\pi/2}^{\pi/2}\frac{e^{i\xi T(\theta)}\operatorname{cs}^{t-2}\theta}{iT(\theta)-y}\,\mathcal{E}\, d\theta$$

And $$G_+ = \frac{1}{2\pi} \int_{-\pi/2}^{\pi/2} \frac{e^{ix\theta} cs^{t-2}\theta}{iT(\theta)-y}\, \mathcal{E}\, d\theta.$$

Similarly,

$$g^-(\xi,t;y) = \int_0^Y e^{s(y-\xi)} (1-2s\cot\varphi - s^2)^{-t/2} ds$$

$$= (\sin^{t-1}\varphi) e^{(\xi-y)\cot\varphi} \int_{\cos\varphi}^{1} e^{(y-\xi)s\csc\varphi} (1-s^2)^{-t/2} ds$$

$$= (\sin^{t-1}\varphi) e^{(\xi-y)\cot\varphi} \int_0^{\varphi} e^{(y-\xi)\csc\varphi\cos\theta} \csc^{t-1}\theta\, d\theta,$$

where we put the upper limit to be the first zero (singularity) of the integrand; this is permissible since we are considering solutions to a recursion of the type $C\, h_{-(n+1)} = h_{-n}$. If we apply C to $\int_0^Y \frac{s^n}{n!} g(x,t;-s)ds$, we get $\int_0^Y \frac{s^n}{n!}(-\frac{\partial}{\partial s}) g\, ds = \frac{s^n}{n!} g\Big|_0^Y + \int_0^Y \frac{s^{n-1}}{(n-1)!} g\, ds$ which yields a zero boundary term as long as g vanishes at Y. g^- is thus an incomplete Bessel function.

And $$G^- = \int_0^Y e^{ys} (1-ks)^{\frac{x-t}{2}} (1+Ks)^{-\frac{x+t}{2}} ds$$

$$= e^{-y\cot\varphi} (\cot^x \varphi/2)(\sin^{t-1}\varphi) \int_{\cos\varphi}^{1} e^{sy\sin\varphi} (1-s)^{\frac{x-t}{2}} (1+s)^{-\frac{x+t}{2}} ds$$

$$= e^{-y\cot\varphi} (\cot^x \varphi/2)(\sin^{t-1}\varphi) \int_0^{\varphi} e^{y\sin\varphi\cos\theta} \tan^x \theta/2\, \sin^{1-t}\theta\, d\theta.$$

5. Integral Representation of J_n. Associated Formula.

$$J_n(x,t) = \frac{n!}{2\pi i} \int_0 (1+kv)^{\frac{t+x}{2}} (1-Kv)^{\frac{t-x}{2}} \frac{dv}{v^{n+1}}$$

$$= \frac{n!}{2\pi i} \int_0 \frac{e^{ixz - tL(iz)}}{V^{n+1}(iz)} \frac{idz}{W(iz)} \quad \text{(we have rotated } z \to iz)$$

$$= \frac{n!}{2\pi i} \int_0 \frac{e^{ixz} cs^{n-t-1} z}{\sin^{n+1} z} \frac{dz}{(i\mathcal{E})^n}.$$

We also have

$$J_n(x,t) = e^{-tL(D)} \frac{1}{W(D)} \left(\frac{D}{V(D)}\right)^{n+1} x^n$$

$$= \csc^n\varphi\, csh^{n-t-1} D \left(\frac{D}{\sinh D}\right)^{n+1} x^n.$$

6. j_n. Recurrence Formula.

By orthogonality,

$$\langle G(x,t;v)^2\rangle = \Sigma \frac{v^{2n}}{n!n!} j_n(t) = e^{tL[2U(v)] - 2tL[U(v)]} = (1+v^2)^t.$$

Thus, $j_n(t) = \langle J_n^2\rangle = n!t^{(n)}$.

We have directly e.g., $J_0 = 1$, $J_1 = \csc\varphi(x + t\cos\varphi)$,

$$J_2 = \csc^2\varphi[x^2+2x(t-1)\cos\varphi+\cos^2\varphi\, t^{(2)}-t].$$

Since $\overline{C}W = \csc\varphi(x\,\mathrm{csh}^2 z - t\,\mathrm{snh} z\ \mathrm{csh} z)$ it is clear that the leading coefficient of J_n is $\csc^n\varphi$. Thus we have the nth degree polynomial

$$J_{n+1} - xJ_n\csc\varphi \perp J_k \quad \text{for } k<n-1.$$

We put $J_{n+1} - x\csc\varphi\, J_n = AJ_n + BJ_{n-1}$. First,

$$\begin{aligned}\langle xJ_nJ_n\rangle &= \frac{j_n}{n!}\langle V^n xJ_n\rangle \\ &= j_n(\langle x\rangle+\langle V'J_1\rangle) \\ &= j_n a(2n-t) \quad \text{(use } \mathrm{csh}D\ x = x-a)\end{aligned}$$

and

$$\begin{aligned}\langle xJ_nJ_{n-1}\rangle &= \frac{j_n}{n!}\langle V^n xJ_{n-1}\rangle \\ &= j_n\langle V'1\rangle \\ &= \varepsilon j_n.\end{aligned}$$

These give us:

$$J_{n+1} = \csc\varphi(x+at-2na)J_n - n(t-n+1)J_{n-1}.$$

7. Kernels. The Number Operator.

The ψ_t function in this case is

$$\overline{\psi}_t(x) = \sum_0^\infty \frac{x^n}{j_n(-t)} = \sum_0^\infty \frac{(-1)^n x^n}{n!\,t_{(n)}} = \Gamma(t)x^{\frac{1-t}{2}} J_{t-1}(2\sqrt{x}),$$

where J is the standard Bessel function, and the bar denotes conjugation $t\to -t$.

By Proposition 6, we have

$$\overline{\psi}_t(x) = \frac{\Gamma(t)}{2\pi i}\int_H e^{s-\frac{x}{s}} s^{-t}\, ds.$$

The associated kernel

$$\begin{aligned}\overline{k}_t(x,y) &= e^{tM}\overline{\psi}_t(y\xi)1 = \frac{\Gamma(t)}{2\pi i}\int_H e^s s^{-t}\overline{G}(x,t;-\tfrac{y}{s})ds \\ &= \frac{\Gamma(t)}{2\pi i}\int_H e^s s^{-t}\left(1-\frac{ky}{s}\right)^{\frac{x-t}{2}}\left(1+\frac{Ky}{s}\right)^{-\frac{x+t}{2}} ds.\end{aligned}$$

The number operator $\overline{C}WV = x\sinh D\ \mathrm{csh}D - t\ \mathrm{snh}D\ \sinh D$

$$= (x+at)\partial_2^- + 2p(x-t)\partial_2^+\partial_2^-,$$

where $\partial_Q^\pm = \dfrac{e^{\pm QD}-1}{\pm Q}$. This is a discrete confluent hypergeometric operator.

The (conjugate) reproducing kernel

$$\overline{K}_t(x,y;\lambda) = e^{tM(x)}e^{tM(y)}\frac{\Gamma(t)}{2\pi i}\int_H e^s s^{-t} e^{-\frac{\lambda\xi\eta}{s}} 1\, ds.$$

These are not very readily calculable in "closed form."

This is the basic discussion for the Bernoulli process. We will follow a similar procedure for the subsequent processes. Also some applications will be indicated with quite surprising variety.

THE SYMMETRIC BERNOULLI PROCESS

Here we consider the process $d(t)$ which is Bernoulli with the probabilities $p,\bar{p}$ both equal to $\frac{1}{2}$. This is thus the process $x(t)$ with $\mathcal{E} = k = K = 1$, $a = 0$, $\varphi = \frac{\pi}{2}$. We can proceed, then, to simply write down the corresponding expressions. We will also see an interesting application of the symmetric theory.

1. $\frac{1}{t}\log\langle e^{zd(t)}\rangle = \log\cosh z$. $d(t)$ is a scaling of $w(t)$ in this case. This is the first limiting case of the general w-process where $\alpha\to 0$. We put $b = -\beta$. Then $w(t) = \sqrt{b/2}\, d(\frac{2t}{b})$, $d(t) = \sqrt{2/b}\, w(\frac{bt}{2})$. $d(t)$ is the same as the process $w(t)$ with $\alpha = 0$, $\beta = -2$. $p_t(x) = 2^{-t}\sum\binom{t}{k}\delta(x-t+2k)$. We have

$$G(x,t;v) = (1+v)^{\frac{t+x}{2}}(1-v)^{\frac{t-x}{2}},$$

$$U(v) = \tfrac{1}{2}\log\frac{1+v}{1-v}, \qquad V(z) = \tanh z$$

$$M(v) = -\tfrac{1}{2}\log(1-v^2), \qquad W(z) = \cosh^2 z$$

$$\xi = x\cosh^2 z, \qquad \overline{C}W = x\cosh^2 z - t\sinh z\cosh z.$$

2. We have $K_n(x,t) = J_n(x,t) = \sum\binom{n}{k}(-1)^{n-k}(\frac{t+x}{2})^{(k)}(\frac{t-x}{2})^{(n-k)}$.

$$\xi^n 1 = \sum\binom{n}{k}(\tfrac{x}{2})^{(k)}(\tfrac{x}{2})_{(n-k)}$$

$$\mu_n(t) = \sum\binom{n}{k}(-1)^k(\tfrac{t}{2})_{(k)}(\tfrac{t}{2})_{(n-k)}.$$

Since $\quad e^{tM(v)} = (1-v^2)^{-t/2}$,

we also have

$$\mu_{2k}(t) = \frac{(2k)!}{k!}(\tfrac{t}{2})_{(k)}, \qquad \mu_{2k+1}(t) = 0, \qquad \mu_n(t) = n!\,i^n C_n^{t/2}(0).$$

The $K_n(x,t)$ are the Krawtchouk polynomials.

3. The canonical density

$$p_t(\xi) = \frac{1}{2\pi}\int_{-\pi/2}^{\pi/2} e^{-i\xi\tan\theta}\cos^{t-2}\theta\, d\theta$$

$$= \frac{1}{\sqrt{\pi}\,\Gamma(t/2)}(\tfrac{\xi}{2})^{\nu}K_\nu(\xi), \quad \text{where } \nu = \frac{t-1}{2}.$$

We then have

$$g_+^- = p_t(y-\xi) = \frac{1}{2\pi}\int_{-\pi/2}^{\pi/2} e^{-iy\tan\theta} e^{i\xi\tan\theta} \cos^{t-2}\theta \, d\theta$$

and
$$G_+^- = \frac{1}{2\pi}\int_{-\pi/2}^{\pi/2} e^{-iy\tan\theta} e^{ix\theta} \cos^{t-2}\theta \, d\theta$$

$$g_+(\xi,t;y) = \frac{1}{2\pi}\int_{-\pi/2}^{\pi/2} \frac{e^{i\xi\tan\theta}\cos^{t-2}\theta}{i\tan\theta - y}\, d\theta$$

$$G_+ = \frac{1}{2\pi}\int_{-\pi/2}^{\pi/2} \frac{e^{ix\theta}\cdot\cos^{t-2}\theta}{i\tan\theta - y}\, d\theta$$

$$g^-(\xi,t;y) = \int_0^{\pi/2} e^{(y-\xi)\cos\theta} \csc^{t-1}\theta \, d\theta$$

$$= \int_0^1 e^{(y-\xi)s}(1-s^2)^{-t/2} ds$$

$$= \sqrt{\pi}\left(\frac{y-\xi}{2}\right)^{\nu}\Gamma\left(1-\frac{t}{2}\right) I_{-\nu}(y-\xi)$$

where $\nu = \frac{t-1}{2}$, I is the standard modified Bessel function.

$$G^- = \int_0^1 e^{ys}(1-s)^{\frac{x-t}{2}}(1+s)^{-\frac{x+t}{2}} ds.$$

The corresponding polynomials can be found explicitly by expanding in power series and integrating.

Also
$$G^- = \int_0^{\pi/2} e^{y\cos\theta} \tan^x \theta/2 \, \sin^{1-t}\theta \, d\theta.$$

4. We can represent K_n as a contour integral

$$K_n(x,t) = \frac{n!}{2\pi i}\oint_0 (1+v)^{\frac{t+x}{2}}(1-v)^{\frac{t-x}{2}} v^{-n-1} dv$$

$$= \frac{n!}{2\pi i}\oint_0 \frac{e^{ixz}\cos^{n-t-1}z}{\sin^{n+1}z}\frac{dz}{i^n}.$$

And following the general theory we have

$$K_n(x,t) = \cosh^{n-t-1} D\left(\frac{D}{\sinh D}\right)^{n+1} x^n.$$

5. The formulae for j_n and recurrence are immediate from the general theory since simply set $\alpha = 0$, $\beta = -2$ to get $j_n(t) = n!t^{(n)}$ and

$$K_{n+1}(x,t) = xK_n - n(t-n+1)K_{n-1}.$$

6. The ψ function here is the same as for the general Bernoulli process:

$$\bar{\psi}_t = \Gamma(t)x^{\frac{1-t}{2}} J_{t-1}(2\sqrt{x}).$$

The number operator is simply $x\partial_2^- + (x-t)\partial_2^+\partial_2^-$. The kernels k and K are, as for the general Bernoulli process, not readily calculable.

7. Application to Residue Calculations.

An interesting application of the theory for $d(t)$ is the answer to the following:

Calculate $\frac{n!}{2\pi i}\int_0 \frac{f(z)}{\sin^{n+1} z}\,dz$, for appropriate analytic f.

We can easily answer this by noting that

$$K_n(x,n-1) = \frac{n!}{2\pi i}\int_0 \frac{e^{ixz}}{\sin^{n+1} z}\frac{dz}{i^n}.$$

Hence, $\frac{n!}{2\pi i}\int_0 \frac{f(z)}{\sin^{n+1} z}\,dz = \frac{n!\,i^n}{2\pi i}\int_0 \frac{e^{Dz}}{\sin^{n+1} z}\frac{dz}{i^n} f(0) = i^n K_n(-iD,n-1)f(0)$.

For example:

$$K_2 = x^2-t, \quad i^2K_2(-iD,1) = D^2+1, \quad \frac{1}{\pi i}\int_0 \frac{f(z)}{\sin^3 z}\,dz = f''(0)+f(0).$$

A direct approach (which is the same as using the alternative representation of K_n) would be to compute the integral as

$$\left(\frac{d}{dz}\right)^n\Big|_0\left(\frac{z}{\sin z}\right)^{n+1} f(z) = \Sigma\binom{n}{k} f^{(k)}(0)\left(\frac{d}{dz}\right)^{n-k}\Big|_0\left(\frac{z}{\sin z}\right)^{n+1}.$$

The terms $\left(\frac{d}{dz}\right)^p\Big|_0\left(\frac{z}{\sin z}\right)^{n+1}$ can be calculated using the expansion of $\frac{z}{\sin z}$ with Bernoulli numbers appearing in the coefficients (equivalently, $\zeta(2n)$ appear as coefficients).

We finally remark that the recursion for K_n gives a recursive solution here, although the expression for general t is calculated first, then t chosen appropriately.

THE EXPONENTIAL PROCESS

The exponential process is obtained from the general w-process by taking the limit $Q \to 0$. It is the dual process, in the sense of time-space duality, to the Poisson process. To see this observe that on functions harmonic with respect to the Poisson process,

$$-\frac{d}{dt} \equiv -T = e^D-1 \quad \text{or} \quad -D = -\log(1-T) \quad \text{the generator for the exponential}$$

process.

1. Define $e(t)$ by $\frac{1}{t}\log\langle e^{ze(t)}\rangle = -\log(1-z)$. $w(t) = -\frac{1}{r}[e(r^2t)-r^2t]$, $e(t) = -rw(r^{-2}t)+t$. Thus, $e(t)$ is the w-process with $Q = 0$, $r = s = -1$, $\alpha = \beta = 2$, which has an added drift of $+t$. $p_t(x) = \frac{x^{t-1}e^{-x}}{\Gamma(t)}\chi(x)$. We have

$$G(x,t;v) = (1+v)^{-t}e^{\frac{v}{1+v}x}$$

$$U(v) = \frac{v}{1+v}, \qquad V(z) = \frac{z}{1-z}, \qquad M(v) = \log(1+v), \qquad W(z) = (1-z)^2$$

$$\xi = x(1-z)^2, \qquad \bar{C}W = (x-tL')W = x(1-z)^2-t(1-z).$$

2. In this case we denote the polynomials, the familiar <u>Laguerre</u> polynomials, by $L_n(x,t)$. We have

$$L_n(x,t) = \Sigma\binom{n}{k}(-1)^{n-k}x^k\,\frac{\Gamma(n+t)}{\Gamma(k+t)}$$

$$\xi^n 1 = \Sigma\binom{n}{k}(-1)^{n-k}x^k\,\frac{\Gamma(n)}{\Gamma(k)}$$

$$\mu_n(t) = t^{(n)}$$

$$e^{tM(v)} = (1+v)^t.$$

3. The canonical density satisfies

$$\int e^{v\xi}p_t(\xi) = (1+v)^t \quad \text{so} \quad p_{-t}(-\xi) = \frac{\xi^{t-1}e^{-\xi}}{\Gamma(t)}\,\chi(\xi).$$

We see that the canonical process is a space-time reflection of the original process.

$$g_+^- = p_t(y-\xi) = \frac{(\xi-y)^{-t-1}e^{y-\xi}}{\Gamma(-t)}\,\chi(\xi-y) = \frac{1}{2\pi}\int e^{is(\xi-y)}(1+is)^t ds$$

$$G_+^- = \frac{1}{2\pi}\int e^{-iys}(1+is)^t e^{\frac{isx}{1+is}}\,ds$$

$$g_+(\xi,t;y) = \frac{1}{2\pi}\int \frac{e^{is\xi}(1+is)^t}{is-y}\,ds, \qquad G_+ = \frac{1}{2\pi}\int \frac{e^{\frac{isx}{1+is}}(1+is)^t}{is-y}\,ds$$

$$g^- = \int_0^1 e^{s(y-\xi)}(1-s)^t ds = \Sigma y^n\Sigma\binom{n+k}{k}(-\xi)^k\,\frac{\Gamma(t+1)}{\Gamma(n+k+t+2)},$$

we use that $\int_0^1 t^{n-1}(1-t)^{m-1}dt$, the beta function, equals $\frac{\Gamma(n)\Gamma(m)}{\Gamma(n+m)}$.

$$G^- = \int_0^1 e^{sy}e^{-\frac{xs}{1-s}}(1-s)^t ds = \int_0^\infty e^{y\frac{s}{1+s}}e^{-sx}(1+s)^{-t-2}ds$$

$$= \sum_0^\infty y^m\Sigma\binom{k+m}{m}\,\frac{(-1)^k(t+2+m)_{(k)}}{x^{k+1+m}}.$$

4. We have the representations

$$L_n(x,t) = \frac{n!}{2\pi i}\int_0 \frac{e^{xz}(1-z)^{t+n-1}}{z^{n+1}}\,dz = (1-D)^{n+t-1}x^n.$$

5. The formula for j_n follows by setting $\beta = 2$ to get

$$j_n(t) = n!t_{(n)}.$$

The recursion is deduced as follows. $e(t)-t$ is the standard $w(t)$ with $\alpha = \beta = 2$. So

$$L_{n+1}(x,t) = (x-t-2n)L_n - n(t+n-1)L_{n-1}.$$

In this case we have a convenient formula for $p_t(x)$ so we can apply GRF to yield the representation

$$L_n(x,t)x^{t-1}e^{-x}\chi(x) = (-1)^n t_{(n)}\left(\frac{D}{1+D}\right)^n x^{t-1}e^{-x}\chi(x).$$

6. The kernels for e(t) can be easily calculated. We have

$$\psi_t(x) = \Sigma\frac{x^n}{n!t_{(n)}} = \Gamma(t)x^{\frac{1-t}{2}}I_{t-1}(2\sqrt{x})$$

$$k_t(x,y) = e^{-tM}\psi_t(y\xi)1 = e^{-tM}\frac{\Gamma(t)}{2\pi i}\int_H e^{y\xi/s}e^s s^{-t}ds\ 1$$

$$= \frac{\Gamma(t)}{2\pi i}\int_H (1+\frac{y}{s})^{-t}e^{\frac{xy}{y+s}}e^s s^{-t}ds \quad (s'+y = s)$$

$$= e^{-y}\psi_t(xy)$$ which is an amazingly simple transformation.

We can proceed to derive the reproducing kernel

$$K_t(x,y;\lambda) = e^{-tM(x)}e^{-tM(y)}\psi_t(\lambda\xi\eta)1$$

$$= e^{-tM(y)}e^{-\lambda\eta}\ \psi_t(\lambda x\eta)1 \quad \text{(by the derivation of } k_t)$$

$$= e^{-tM(y)}\int_H e^{(\frac{\lambda x}{s}-\lambda)}e^s s^{-t}ds\ 1$$

$$= (1-\lambda)^{-t}\exp[-(x+y)\frac{\lambda}{1-\lambda}]\psi_t(xy\frac{\lambda}{(1-\lambda)^2})$$

$$= G(x+y,t;-\lambda)\psi_t(\frac{xy\lambda}{(1-\lambda)^2})$$

$$= \Sigma\frac{\lambda^n L_n(x,t)L_n(y,t)}{n!t_{(n)}}.$$

This gives us an exponential transform,

$$f^e(\lambda,x) = \lambda^{\overline{C}WV}f(x) = (1-\lambda)^{-t}e^{-\frac{x\lambda}{1-\lambda}}\int_0^\infty f(y)e^{-\frac{y}{1-\lambda}}\psi_t(\frac{xy\lambda}{(1-\lambda)^2})y^{t-1}\frac{dy}{\Gamma(t)}.$$

If we choose t=1 we have

$$f(x) = \lim_{\lambda\to 1} f^e(\lambda,x) = \lim_{\lambda\to 1}(1-\lambda)^{-1}e^{-\frac{x\lambda}{1-\lambda}}\int_0^\infty f(y)e^{-\frac{y}{1-\lambda}}I_0(\frac{2}{1-\lambda}\sqrt{xy\lambda})dy.$$

Finally, we note that the number operator $\overline{C}WV = (x-t)\frac{\partial}{\partial x} - x\frac{\partial^2}{\partial x^2}$ is just minus the confluent hypergeometric operator.

7. The Gamma Transform. Inversion.

The expansion theorem gives us

$$e^{-tL}f(a+x) = \sum_0^\infty \frac{L_n(x,t)}{n!} V^n f(a).$$ We have then,

$$\langle f[e(t)]\rangle = e^{tL}f(0) = \sum_0^\infty \frac{L_n(-t,-t)}{n!} V^n f(t).$$

Define $e_n(t) = L_n(-t,-t)$. $e_0(t) = 1$, $e_1(t) = 0$, $e_2(t) = t$, $e_3(t) = -4t$, $e_4(t) = 3t^2+18t$.

From the recursion for L_n we have $e_{n+1}(t) = -2ne_n + n(t+1-n)e_{n-1}$.

Thus, $\deg e_{2k} = \deg e_{2k+1} = k$.

We define the gamma transform, a generalized Laplace transform, by

$$\Gamma_t f(\lambda) = \langle f(\frac{e(t)}{\lambda})\rangle = \int_0^\infty f(y) \frac{\lambda^t y^{t-1} e^{-\lambda y}}{\Gamma(t)} dy.$$

t=1 is the ordinary Laplace transform. An example for $\lambda=1$ is

$$\zeta(t) = \Gamma_t f(1) \quad \text{where} \quad f(x) = (1-e^{-x})^{-1}.$$

The expansion theorem yields (scaling by λ appropriately)

$$\Gamma_t f(\lambda) = \sum_0^\infty \frac{e_n(t)}{n!} (\frac{D}{\lambda-D})^n f|_{t/\lambda}$$

$$= f(\frac{t}{\lambda}) + \sum_2^\infty \frac{e_n(t)}{n!} (\frac{D}{\lambda-D})^n f|_{t/\lambda}.$$

$(\frac{D}{\lambda-D})^n$ can be calculated either as a series $\sum_0^\infty \frac{n_{(k)}}{k!} \frac{D^{k+n}}{\lambda^{k+n}}$ or by the integral

$$(\frac{D}{\lambda-D})^n f(x) = \int_0^\infty f^{(n)}(x+y) \frac{y^{n-1}e^{-\lambda y}dy}{\Gamma(n)}$$

which can be seen by noting that

$$(\frac{\lambda}{\lambda-D})^n f(x) = \int_0^\infty e^{yD} \frac{\lambda^n y^{n-1} e^{-\lambda y}}{\Gamma(n)} f(x).$$

We first observe that for fixed t,

$$\lim_{\lambda\to\infty} \Gamma_t f(\lambda) = f(0).$$

This follows since the terms for $n \geq 2$ have a factor λ^{-n}.

Secondly we have the

<u>Inversion Formula</u>.

$$\lim_{\lambda\to\infty} \Gamma_{x\lambda} f(\lambda) = f(x).$$

<u>Proof</u>: Since $\deg(e_n) \leq \frac{n}{2}$, the terms for $n \geq 2$ are of weight $\leq (x\lambda)^{n/2}\lambda^{-n}$ which vanish as $\lambda\to\infty$.

This gives us the usual "real" inversion formula for Laplace transforms. Observe that

$$\Gamma_n f(x) = \int_0^\infty f(y)\,\frac{x^n y^{n-1} e^{-xy}}{\Gamma(n)}\,dy$$

$$= \frac{x^n(-D)^{n-1}}{\Gamma(n)} \int_0^\infty f(y)e^{-xy}dy$$

$$= \frac{x^n(-D)^{n-1}}{\Gamma(n)} \mathcal{L}f(x),$$

$\mathcal{L}$ denoting Laplace transform.

Putting $x\lambda = n$, we have

$$f(x) = \lim_{n\to\infty} \Gamma_n f(\tfrac{n}{x}) = \lim_{n\to\infty} (\tfrac{n}{x})^n \frac{(-D)^{n-1}}{\Gamma(n)} \mathcal{L}f(\tfrac{n}{x}).$$

THE POISSON PROCESS

The Poisson process is the standard model of a jump process that increases by one unit at a time with an exponentially distributed waiting time between jumps. Just as the exponential process is the "time" process for the Poisson process, the Poisson is the "time" process for the exponential process. It arises in our context as the limiting case $\beta \to 0$, i.e. π is linear.

1. Define $N(t)$ by $\frac{1}{t}\log\langle e^{zN(t)}\rangle = e^z - 1$. Then $w(t) = \alpha[N(\frac{t}{\alpha^2}) - \frac{t}{\alpha^2}]$, $N(t) = \frac{1}{\alpha}w(\alpha^2 t) + t$. So $N(t)-t$ is the w-process with $\beta = 0$, $Q = \alpha = 1$, $r = -1$. $p_t(x) = e^{-t}\sum_0^\infty \frac{t^k}{k!}\,\delta(x-k)$. We have

$$G(x,t;v) = e^{-vt}(1+v)^x$$

$$U(v) = \log(1+v), \qquad V(z) = e^z - 1$$

$$M(v) = v, \qquad W(z) = e^{-z}$$

$$\xi = xe^{-z}, \qquad \overline{C}W = (x-tL')W = xe^{-z} - t = \xi - t.$$

2. In this case the polynomials are the <u>Poisson-Charlier</u> polynomials which we denote by $P_n(x,t)$. We have

$$P_n(x,t) = \Sigma\binom{n}{k}(-t)^k x^{(n-k)}$$

$$\xi^n 1 = x(x-1)\ldots(x-n+1) = x^{(n)}$$

$$\mu_n(t) = t^n$$

$$e^{tM(v)} = e^{tv}.$$

3. The canonical process is this case is not even random! Since the generator is just V, we have $\xi(t) = \xi+t$. The canonical density is, of course, $p_t(\xi) = \delta(\xi-t)$.

$$g_+^- = p_t(y-\xi) = \delta(\xi+t-y)$$

$$G_+^- = \frac{1}{2\pi}\int e^{ist}e^{-isy}e^{is\xi_1}\,ds$$

$$= \frac{1}{2\pi}\int e^{ist}e^{-isy}(1+is)^x\,ds$$

$$= \frac{1}{2\pi}\int e^{is(y-t)}(1-is)^x\,ds$$

$$= \frac{(t-y)^{-x-1}e^{y-t}}{\Gamma(-x)}\,\chi(t-y).$$

In particular,

$$p_t(-\xi)1 = \frac{t^{-x-1}e^{-t}}{\Gamma(-x)}\,\chi(t),$$

the distribution for an exponential process at "time" -x.

$$g_+(\xi,t;y) = \int_0^\infty e^{ys}\delta(\xi+t-s)ds = e^{y(t+\xi)}\chi(\xi+t)$$

$$G_+ = \int_0^\infty e^{ys}G_+^-(x,t;s)ds = \int_0^t e^{ys}(t-s)^{-x-1}e^{s-t}ds/\Gamma(-x)$$

$$= \Sigma\,\frac{y^n}{n!}\int_0^t s^n(t-s)^{-x-1}e^{s-t}ds/\Gamma(-x)$$

$$= \Sigma\,\frac{y^n}{n!}\,e^{-t}\,\Sigma\,\frac{t^k}{k!}\,t^{n-x}\,\frac{\Gamma(n+k+1)}{\Gamma(n+k+1-x)} \qquad \text{(via the beta function)}$$

$$= \Sigma\,\frac{y^n}{n!}\,t^{n-x}\langle\frac{1}{[n+1+N(t)]_{(-x)}}\rangle.$$

Similarly,

$$g^- = \int_0^\infty e^{s(y-\xi)}e^{-st}ds = (\xi+t-y)^{-1} = \Sigma\, y^n(\xi+t)^{-n-1}$$

and

$$G^- = \int_0^1 e^{sy}e^{-st}(1-s)^x ds = \Sigma\, y^n e^{-t}\,\Sigma\,\frac{t^k}{k!}\,\frac{\Gamma(k+x+1)}{\Gamma(k+x+n+2)}$$

$$= \Sigma\, y^n\langle\frac{1}{[x+N(t)+1]_{(n+1)}}\rangle.$$

4. We can represent

$$P_n(x,t) = \frac{n!}{2\pi i}\oint_0 e^{-vt}(1+v)^x v^{-n-1}dv$$

$$= \frac{n!}{2\pi i}\oint_0 e^{z(x+1)-t(e^z-1)}(e^z-1)^{-n-1}dz$$

$$= e^{-t(e^D-1)}\left(\frac{D}{e^D-1}\right)^{n+1}(x+1)^n.$$

The series $\frac{D}{e^D-1}$ can be expanded using Bernoulli numbers.

5. The formula for j_n with $\beta=0$ yields $j_n(t) = n!t^n$. And the recursion formula follows by putting $N(t)-t$ for the w-process with $\alpha=1$, $\beta=0$. Thus,

$$P_{n+1}(x,t) = (x-t-n)P_n-ntP_{n-1}.$$

We can use GRF to obtain another representation

$$P_n(x,t)p_t(x) = t^n(e^{-z}-1)^n p_t(x) = (-1)^n t^n (\partial_1^-)^n p_t(x).$$

6. The ψ function reduces since $\beta=0$. We have

$$\psi_t(x) = \Sigma \frac{x^n}{n!t^n} = e^{x/t}.$$

This shows the asymptotic relation, putting $n = \frac{2}{\beta}$,

$$\lim_{n\to\infty} \Gamma(nt)(nx)^{\frac{1-nt}{2}} I_{nt-1}(2\sqrt{nx}) = e^{x/t}.$$

The associated kernel is readily found:

$$k_t(x,y) = e^{-tM}e^{y\xi/t}1 = G(x,\tfrac{y}{t};t) = e^{-y}(1+\tfrac{y}{t})^x.$$

However, the reproducing kernel is not readily evaluated. We have, for example,

$$K_t(x,y;\lambda) = e^{-tM(x)}e^{-tM(y)}e^{\lambda\xi\eta/t}1$$

$$= e^{-tM(y)}e^{-\lambda\eta}(1+\tfrac{\lambda\eta}{t})^x 1.$$

To find an explicit form for K_t we proceed as follows.

$$(-1)^n P_n(x,t) = \Sigma\binom{n}{k}(-1)^k t^{n-k} x^{(k)} = \Sigma \frac{x^{(k)}}{k!}(-1)^k T^k t^n = (1-T)^x t^n,$$

where $T = \frac{d}{dt}$. Thus,

$$P_n(x,t) = (-1)^n L_n(t,x-n+1),$$

another example of the Poisson-Exponential duality. We have

$$K_t(x,y;\lambda) = \sum_0^\infty \frac{\lambda^n P_n(x,t)P_n(y,t)}{n!t^n} = (1-T)^x(1-U)^y e^{atu}\Big|_{\substack{a=\lambda/t\\u=t}},$$

where $U = \frac{d}{du}$.

We use the formulas $g(D)^o e^{ax} = e^{ax}g(a+D)$ which follows from EL and Prop. 1 or directly from GLM (also see Prop. 8); and $g(\lambda D) = \lambda^{-xD}g(D)\lambda^{xD}$ (see Prop. 15 and preceding). We have, then, first applying EL,

$$(1-T)^x(1-U)^y e^{atu} = (1-T)^x(1-at)^y e^{atu} = e^{atu}(1-T-au)^x(1-at)^y.$$

Let $1-at' = t$; then $T' = -aT$. Then $(1-T'-au)^x(1-at')^y$ becomes

$$(1+aT-au)^x t^y = (1-au)^x(1-\frac{-a}{1-au}T)^x t^y = (1-au)^x(\frac{1-au}{-a})^{tT}(1-T)^x(\frac{-a}{1-au})^y t^y$$

$$= (1-au)^x(\frac{-a}{1-au})^y(-1)^y P_y[x,t(\frac{1-au}{-a})].$$

Thus,

$$K_t(x,y) = e^{atu}(1-au)^{x-y}a^y P_y[x,(1-at)(\frac{1-au}{-a})]\Big|_{\substack{a=\lambda/t\\ u=t}}$$

$$= e^{\lambda t}(1-\lambda)^{x-y}(\frac{\lambda}{t})^y P_y[x,-\lambda^{-1}(1-\lambda)^2 t].$$

$P_y(x,t) = (-1)^y L_y(t,x-y+1)$ can be expressed for non-integral y, e.g. by using integral forms of P_n or L_n and analytically continuing $n\to y$, or using standard confluent hypergeometric functions.

The number operator $\overline{C}WV$ is simply $(x-t)\partial_1^- - t\partial_1^+\partial_1^-$.

7. Discrete Calculus.

It is clear that the calculus of the Poisson process is discrete, or difference, calculus. The operator $V = e^z-1 = \partial_1^+$. Thus, the expansion theorem yields

$$f(x) = \sum_0^\infty \frac{P_n(x,t)}{n!}\langle\partial_1^{+n} f[N(t)]\rangle.$$

We can use the inversion principle to see that

$$x^{(n)} = \Sigma\binom{n}{k}t^k P_{n-k}(x,t) \quad \text{and hence} \quad \langle N^{(n)}\rangle = t^n.$$

Since a Poisson process jumps one step at a time we have

$$\int_0^t f[N(s)]dN(s) = \sum_0^{N(t)} f(j)$$

and $$\int_0^\infty f[N(s)]dN(s) = \sum_0^\infty f(j), \quad \text{since } N(t)\to\infty \text{ a.s. as } t\to\infty.$$

E.g., $$\int_0^t N^{(n)}dN = \frac{N^{(n+1)}}{n+1}, \quad VN^{(n)} = nN^{(n-1)}.$$

So V is the inverse to integration with respect to the (random) measure dN.

We also have, e.g.

$$\langle N^{(n)}P_k[N(t),t]\rangle = t^k\langle V^k N^{(n)}\rangle = n^{(k)}t^k\langle N^{(n-k)}\rangle = n^{(k)}t^n.$$

The P_n's are a natural orthogonal system for discrete analysis.

Examples.

1. Express x^n in terms of $x^{(k)}$'s.

The expansion theorem $f(x+a) = \Sigma \frac{P_k(x,t)}{k!} e^{tL}V^k f(a)$, with $t=0$, yields

$$f(x+a) = \Sigma \frac{x^{(k)}}{k!} V^k f(a).$$

We compute

$$V^k a^n = (e^z-1)^k a^n = \Sigma \binom{k}{\ell}(-1)^{k-\ell}(a+\ell)^n.$$

Setting $c_{nk} = \frac{1}{k!}\Sigma\binom{k}{\ell}(-1)^{k-\ell}\ell^n$ we have $x^n = \Sigma c_{nk} x^{(k)}$.

2. Application to the operator xD.

From $\lambda^{xD} f(x) = f(x\lambda)$ we get $(xD)^{(k)} f(x) = \partial_\lambda^k\big|_{\lambda=1}\lambda^{xD} f(x) = x^k D^k f(x)$.

Using (1) we thus have $(xD)^n = \Sigma c_{nk} x^k D^k$.

BROWNIAN MOTION

We turn to our final standard process, Brownian motion, $b(t)$. This is the extreme limiting case of $w(t)$, $\alpha\to 0$, $\beta\to 0$. If we take the further limiting case, $t\to 0$, we recover the original powers x^n and we see Taylor series as a special orthogonal expansion.

1. We have $\frac{1}{t}\log\langle e^{zb(t)}\rangle = \frac{z^2}{2}$. $b(t) = w(t)$ with $\alpha = \beta = 0$. One can scale $b(t) = \sqrt{t}\,b(1)$, in distribution; but we generally deal with $b(t)$ directly.

$$p_t(x) = \frac{1}{\sqrt{2\pi t}} e^{-x^2/2t}.$$

We have $G(x,t;v) = e^{vx-v^2t/2}$

$U(v) = v,$	$V(z) = z$
$M(v) = \frac{v^2}{2},$	$W(z) = 1$
$\xi = x,$	$\overline{C}W = x-tz.$

2. The J_n's are the Hermite polynomials $H_n(x,t)$. Here

$$H_n(x,t) = \Sigma\binom{n}{2k}(-t)^k x^{n-2k}\frac{(2k)!}{2^k k!}$$

$$\xi^n 1 = x^n$$

$$\mu_{2k}(t) = \frac{(2k)!}{2^k\, k!}\, t^k, \qquad \mu_{2k+1}(t) = 0$$

$$e^{tM(v)} = e^{v^2 t/2}.$$

3. Here the canonical process is identical with the original process. Thus the g's and G's are the same.

$$g_+^- = p_t(y-x) = \frac{1}{\sqrt{2\pi t}}\, e^{-(x-y)^2/2t}$$

$$g_+(x,t;y) = \int_0^\infty e^{sy} p_t(s-x)ds = e^{yx+y^2t/2}\Phi(-\frac{x+yt}{\sqrt{t}}),$$

where $\Phi(x) = \int_x^\infty e^{-s^2/2}\, ds.$

$$\bar{g} = \int_0^\infty e^{sy} e^{-xs-s^2t/2}\, ds = \sqrt{2\pi/t}\; e^{(x-y)^2/2t}\Phi(\frac{x-y}{\sqrt{t}}).$$

4. We have $H_n(x,t) = \frac{n!}{2\pi i} \int_0 e^{vx-v^2t/2} v^{-n-1} dv$

$$= e^{-tD^2/2} x^n.$$

Observe that the H_n's are the moment polynomials $\bar{h}_n(x,t)$ in this case since $V(D) = D$, $W(D) = 1$.

5. The formula for j_n with $\beta = 0$ gives $j_n(t) = n!t^n$.
We can use the recursion formula directly to yield

$$H_{n+1}(x,t) = xH_n - ntH_{n-1}$$

And GRF applies to give the representation

$$H_n(x,t) = e^{x^2/2t}(-tD)^n e^{-x^2/2t}.$$

6. The ψ, k, K functions can be computed readily in this case.

$$\psi_t(x) = \Sigma \frac{x^n}{n!t^n} = e^{x/t}$$

$$k_t(x,y) = e^{-tM}\psi(y\xi)1 = e^{-tL} e^{xy/t} = e^{(xy/t)-(y^2/2t)}.$$

$K_t(x,y)$ is most readily computed as follows. Observe that in this case the number operator $\overline{CWV}$ is the same as $\bar{A}$, namely

$$\overline{CWV} = \bar{A} = xD-tD^2 \quad \text{or} \quad x\frac{\partial}{\partial x} - t\frac{\partial^2}{\partial x^2}.$$

So

$$K_t(x,y;\lambda)p_t(y) = \lambda^{\bar{A}}\delta(x-y)$$

$$= e^{-tL}\lambda^{xD}e^{tL}\delta(x-y)$$

$$= \lambda^{xD}e^{-t[L(\lambda D)-L(D)]}\delta(x-y)$$

$$= \lambda^{xD}e^{t(1-\lambda^2)D^2/2}\delta(x-y)$$

$$= p_{t(1-\lambda^2)}(y-\lambda x).$$

$$\therefore\ K_t(x,y;\lambda) = \frac{e^{y^2/2t}}{\sqrt{1-\lambda^2}}\exp\left[\frac{1}{2t(\lambda^2-1)}(y^2-2\lambda xy+\lambda^2x^2)\right]$$

$$= (1-\lambda^2)^{-\frac{1}{2}}\exp\left[\frac{1}{2t(\lambda^2-1)}(\lambda^2x^2+\lambda^2y^2-2\lambda xy)\right],$$

which is Mehler's kernel.

CANONICAL MOMENTS

In this section we consider the moments $\mu_n(t)$ of $p_t(\xi)$. We recall the section on space-time duality in Chapter III where we defined the characteristic operators $\ell = L'{}^{\circ}L^{-1}$, $m = \ell^{-1} = L^{\circ}L'^{-1}$. We have, for $T = d/dt$,

$$e^{tM(v)} = \Sigma\frac{v^n}{n!}\mu_n(t)$$

$$M^{-1}(T)\mu_n(t) = n\mu_{n-1}$$

$$tM'{}^{\circ}M^{-1}(T)\mu_n(t) = \mu_{n+1}.$$

In terms of L, $M = L^{\circ}U$, $M^{-1} = V^{\circ}L^{-1}$, $M'{}^{\circ}M^{-1} = \ell\cdot W^{\circ}L^{-1}$. In the general theory, $M = m$, but, as we have seen, the canonical L's have been sometimes uncentered so $V \neq L'$. However, we have, for the general theory, $M^{-1} = \ell$, $M'{}^{\circ}M^{-1} = \frac{2\ell}{\pi(\ell)}$. Here we denote $M^{-1}(T)$ by m^- and $M'{}^{\circ}M^{-1}$ by m^+. The number operator for the μ_n's is thus tm^+m^-. Using our earlier results let's evaluate m^-,m^+ for our standard processes:

	$M(v)$	$\mu_n(t)$	m^+	m^-
Bernoulli	$-\frac{1}{2}\log(1+2v\cot\varphi-v^2)$	$n!i^nC_n^{t/2}(i\cot\varphi)$	$e^{2T}\sqrt{\mathcal{E}^{-2}-e^{-2T}}$	$\cot\varphi+\sqrt{\mathcal{E}^{-2}-e^{-2T}}$
Symmetric Bernoulli	$-\frac{1}{2}\log(1-v^2)$	$\frac{2k!}{k!}(t/2)_{(k)}$	$e^{2T}\sqrt{1-e^{-2T}}$	$\sqrt{1-e^{-2T}}$
Exponential	$\log(1+v)$	$t^{(n)}$	e^{-T}	e^T-1
Poisson	v	t^n	1	T
Brownian Motion	$v^2/2$	$\frac{2k!}{2^kk!}t^k$	$\sqrt{2T}$	$\sqrt{2T}$

Remarks: (1) In the Bernoulli case we note that $(tm^+)^2\mu_{2k}(t) = \mu_{2k+2}(t)$ and $(tm^+)^2 = \varepsilon^{-2}\tau^2-(t+1)\tau$, $\tau = te^{2T}$.

(2) In the symmetric processes, Bernoulli and Brownian motion, the odd moments are zero so the determination of the square root is, in this context at any rate, not an essential problem. For the general Bernoulli we choose the + root in M^{-1} to agree with the symmetric case when $\varphi \to \pi/2$.

SCALINGS. SUMMARY OF FORMS.

We summarize here the various scalings used to standardize our processes. In the next chapter we will use these scalings to determine the interrelationships among our processes. We give explicitly the "full" unscaled L, V, W, G, M for our five processes.

1. Bernoulli: $v = \frac{\varepsilon}{2q}v' = \sqrt{-(\beta/2)}\,v'$. $t = -\frac{2t'}{\beta}$. $x + at = 2qx'$.

We have $\frac{d}{dx} = \frac{1}{2q}\frac{d}{dx'}$. And $L = -\frac{\alpha}{\beta}z - \frac{2}{\beta}\log \operatorname{csh}Qz/2$.

$$U = q\log\frac{1-(v/r)}{1-(v/s)}, \qquad V = \frac{s\partial_Q^+}{s+\partial_Q^+} = 2q\frac{\sinh Qz/2}{\operatorname{csh}Qz/2}, \qquad W = \operatorname{csh}^2 Qz/2.$$

$$G = (1-\tfrac{v}{r})^{qx-qrt}(1-\tfrac{v}{s})^{qst-qx}. \qquad M(v) = q\log[(1-\tfrac{v}{r})^r(1-\tfrac{v}{s})^{-s}].$$

$$\overline{C}WV = x\partial_Q^- + (\tfrac{x}{s}-t)\partial_Q^+\partial_Q^-.$$

2. Symmetric Bernoulli: $b \equiv -\beta$. $t = \frac{2t'}{b}$. $x = \sqrt{2/b}\,x'$. $L = \frac{2}{b}\log\cosh\sqrt{b/2}\,z$.

$$U = \frac{1}{\sqrt{2b}}\log\frac{1+\sqrt{b/2}\,v}{1-\sqrt{b/2}\,v}, \qquad V = \sqrt{2/b}\tanh\sqrt{b/2}\,z, \qquad W = \cosh^2\sqrt{b/2}\,z.$$

$$G = (1+\sqrt{b/2}\,v)^{(t/b)+(x/\sqrt{2b})}(1-\sqrt{b/2}\,v)^{(t/b)-(x/\sqrt{2b})}.$$

$$M(v) = -\frac{1}{b}\log(1-\frac{bv^2}{2}). \qquad \overline{C}WV = x\partial_Q^- + (\tfrac{x}{s}-t)\partial_Q^+\partial_Q^-.$$

We have $v = \sqrt{b/2}\,v'$ for canonical scaling.

3. Exponential: $t = r^2t'$. $x-t = -rx'$.

$$L = rz-r^2\log(1+\tfrac{z}{r}).$$

$$U = \frac{rv}{r-v}, \qquad V = \frac{rz}{r+z}, \qquad W = (1+\tfrac{z}{r})^2. \qquad G = (\tfrac{r}{r-v})^{r^2t}\exp[r\tfrac{v}{r-v}(x-rt)].$$

$$M(v) = \frac{r^2v}{r-v}-r^2\log\frac{r}{r-v}. \qquad \overline{C}WV = x\frac{\partial}{\partial x} + (\tfrac{x}{r}-t)\frac{\partial^2}{\partial x^2}.$$

$\frac{rz}{r+z}$ becomes, via the scaling, $x-t = -rx'$, $\frac{d}{dx} = -\frac{1}{r}\frac{d}{dx'}$, $\frac{-r^2 z}{r-rz} = -r\frac{z}{1-z}$.

That is, $v = -r^{-1}v'$, since the canonical $V = \frac{z}{1-z}$.

4. Poisson: $t = \alpha^{-2}t'$. $x-t = \alpha^{-1}x'$.

$$L = \alpha^{-2}(e^{\alpha z}-1-\alpha z).$$

$$U = \frac{1}{\alpha}\log(1+\alpha v), \quad V = \partial_\alpha^+, \quad W = e^{-\alpha z}. \quad G = (1+\alpha v)^{x/\alpha+t/\alpha^2}e^{(-1/\alpha)vt}.$$

$$M(v) = \frac{v}{\alpha}-\frac{1}{\alpha^2}\log(1+\alpha v). \quad \bar{C}WV = x\partial_\alpha^- - t\partial_\alpha^+\partial_\alpha^-.$$

For standard Poisson, ∂_α^+ becomes $\alpha^{-1}\partial_1^+$, i.e. $v = \alpha v$!

5. Brownian motion: $L = M = \frac{z^2}{2}$, $U = v$, $V = z$, $W = 1$, directly.

$$G = e^{vx-(v^2t/2)}. \quad \bar{C}WV = x\frac{\partial}{\partial x} - t\frac{\partial^2}{\partial x^2}.$$

We have $b(t) = \sqrt{t}\, b(1)$ in distribution.

This yields a scaling $t = 1$, $x = t^{-\frac{1}{2}}x'$, $v = t^{\frac{1}{2}}v'$.

LAGRANGIANS

We will consider the full L's we have as Hamiltonians and compute the corresponding Lagrangians. We observe that $R'(\alpha) = L'^{-1}(\alpha) = U(\alpha)$ so that $R(\alpha) = \int U(\alpha) + \text{constant}$. The relation $H[R'(\alpha)] = \alpha R'(\alpha)-R(\alpha)$ becomes $M(\alpha) = \alpha U(\alpha)-R(\alpha)$. Setting $\alpha = 0$ yields $R(0) = 0$. Thus,

$$R(\alpha) = \int_0^\alpha U(\alpha)d\alpha = \alpha U(\alpha)-M(\alpha).$$

The momentum is $U(\dot{x})$.

Remark: $R(\alpha) = \log G(\alpha,1;\alpha)$.

1. We start with the symmetric Bernoulli process. Set $b = 2k^2$.

$$L = k^{-2}\log \cosh kz. \qquad U = \frac{1}{2k}\log\frac{1+kv}{1-kv}.$$

$$M(v) = -\frac{1}{2k^2}\log(1-k^2v^2).$$

Thus, $$R(\alpha) = \left(\frac{1}{2k^2}+\frac{\alpha}{2k}\right)\log(1+k\alpha)+\left(\frac{1}{2k^2}-\frac{\alpha}{2k}\right)\log(1-k\alpha).$$

2. Exponential process. $L = rz-r^2\log(1+r^{-1}z)$. $U = \frac{rv}{r-v}$.

$$M(v) = \frac{r^2v}{r-v} - r^2\log\frac{r}{r-v}.$$

Thus, $$R(\alpha) = -r\alpha + r^2\log\frac{r}{r-\alpha}.$$

3. Poisson process. $L = \alpha^{-2}(e^{\alpha z}-1-\alpha z)$. $U = \alpha^{-1}\log(1+\alpha v)$.

$$M(v) = \alpha^{-1}v - \alpha^{-2}\log(1+\alpha v).$$

Here we use $\dot{x}$ for the argument of R.

$$R(\dot{x}) = -\alpha^{-1}\dot{x} + \alpha^{-2}(1+\alpha\dot{x})\log(1+\alpha\dot{x}).$$

4. Brownian motion. $L = z^2/2$. $U = v$.

$M = v^2/2$. $R(\alpha) = \alpha^2/2$.

This corresponds physically to ordinary free motion of a particle with mass 1.

Remarks:

(1) It is an interesting problem to explore the relationships among the classical, quantum and probabilistic processes associated with a given Lagrangian.

(2) Since the Hamiltonian $L(z)$ is independent of x, $\dot{z}(t) = 0$, so $z(t) = z$ is the momentum. Observe that $V(z) = V[U(\alpha)] = \alpha = \dot{x}$ is the speed of travel in terms of the momentum. $M(\dot{x})$ is the energy expressed in terms of $\dot{x}$.

PART 2. FURTHER PROBABILISTIC ASPECTS

CHAPTER VI. LIMIT THEOREMS

We will proceed to a detailed study of the connections among our standard processes. We have the basic picture

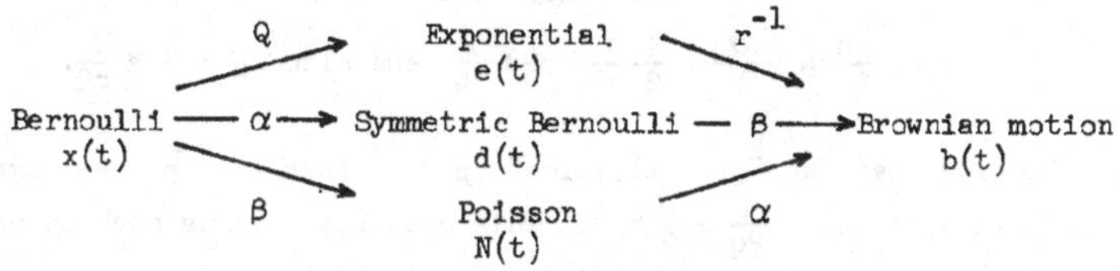

where the limit is taken by the vanishing of the indicated variable. The limits indicate convergence of the corresponding generators, thus they give convergence in distribution. We proceed to examine the limit theorems corresponding to our picture.

BERNOULLI → EXPONENTIAL

1. The generator is $L = -\frac{\alpha}{\beta}z - \frac{2}{\beta}\log\operatorname{csh}\frac{Qz}{2}$, $\quad V = \frac{s\partial_Q^+}{s+\partial_Q^+}$.

   The process $w_Q(t) = \frac{1}{2q}x(-\frac{2t}{\beta}) - \frac{\alpha t}{\beta}$, where $x(t)$ is the standard Bernoulli process.

   As $Q \to 0$, $r' \approx s'$ and we must specify $r', s' \to r$.

   Since $r = \frac{-\alpha+Q}{\beta}$, in the limit $r = -\frac{\alpha}{\beta} = \frac{-2\alpha}{\alpha^2} = -\frac{2}{\alpha} = -\sqrt{2/\beta}$.

   Thus, the limit is determined by $\alpha^2 \approx 2\beta$, $-\alpha \approx \beta r$.

   Clearly then,

$$L \approx rz - r^2\log(\operatorname*{lt}_{Q\to 0}\operatorname{csh}\frac{Qz}{2}).$$

   We have $\operatorname*{lt}_{Q\to 0}\operatorname{csh}\frac{Qz}{2} = \operatorname*{lt}_{Q\to 0} p(e^{Qz}-1)+1 = \operatorname*{lt}_{Q\to 0}\frac{Q-\alpha}{2Q}(e^{Qz}-1)+1 = \frac{z}{r}+1.$

$$\operatorname*{lt}_{Q\to 0} V_Q(z) = \frac{sz}{s+z} = \frac{rz}{r+z}.$$

   In terms of processes, then, we have

<u>Proposition 29</u>: (1) The process $\frac{1}{2q}x_Q(-r^2t)$, where x_Q is a Bernoulli process with "Prob" (jump of $+Q$) $= p = \frac{1}{rQ}$, converges as $Q \to 0$ to the modified exponential process $-r^{-1}e(r^2t)$.

(2) In particular, choosing $p = -\frac{1}{Q}$, the process $\frac{1}{2q}x_Q(-t)$

converges to the exponential process $e(t)$ as $Q \to 0$.

<u>Proof</u>: $w_Q(t) = \frac{1}{2q}x_Q(-\frac{2t}{\beta})-\frac{\alpha t}{\beta}$. Using the asymptotic relations for α and β we have $\frac{1}{2q}x_Q(-r^2t)+rt \underset{Q\to 0}{\to} -\frac{1}{r}[e(r^2t)-r^2t]$ since this is $w(t)$ corresponding to the limiting L. Now we note that

$$p = \frac{1-a}{2} = \frac{1-\alpha q}{2} = \frac{1}{2}\frac{\alpha}{2Q} \approx \frac{1}{2}+\frac{1}{rQ} \quad \text{and also} \quad p = \frac{q}{s} \approx \frac{1}{rQ}.$$

<u>Remarks</u>: Observe that the "probabilities" $p \to +\infty$ (sgn r), $\bar{p} \to -\infty$ (sgn r). The coefficient $\frac{1}{2q} = \frac{Q}{2} \to 0$. This does not give us much of an idea of what the process $e(t)$ is phenomenologically.

2. For the J_n's we have

<u>Proposition 30</u>: $\underset{q\to\infty}{\ell t}\, \Sigma\binom{n}{k} r^{k-n}s^{-k}(qx-qrt)^{(n-k)}(qst-qx)^{(k)} = (-1)^n r^{-n}L_n(r^2t-rx,r^2t)$, where L_n are the Laguerre polynomials.

<u>Proof</u>: $e(t)$ is derived from $w(t)$ by the scaling $t = r^2t'$, $x-t = -rx'$, $v = -r^{-1}v'$.
From $r^2t-rw(t) = e(r^2t)$ we have that $J_n(x,t)$ for the corresponding $w(t)$ process is $(-r)^{-n}L_n(r^2t-rx,r^2t)$.

<u>Examples</u>: $J_1 : x \to -r^{-1}(r^2t-rx-r^2t) = x$

$J_2 : x^2-\alpha x-t \to x^2+\frac{2}{r}x-t$

$$L_2(x,t) = x^2 - 2x(t+1) + t(t+1) \to r^{-2}(r^2t-rx)^2 + 2(rx-r^2t)(r^2t+1) + r^2t(r^2t+1)$$

reduces correctly.

3. We also have convergence of the moments. From the scalings $t = -\frac{2t'}{\beta}$ and $v = \sqrt{-\beta/2}\, v'$, we have the moments for $x(t)$,

$$\mu_n(t) = n!i^n(-\frac{\beta}{2})^{n/2}C_n^{-t/\beta}(i\cot\varphi).$$

As $Q \to 0$, $\beta \approx 2r^{-2}$; $\cot\varphi = \frac{\alpha q}{\sqrt{1-\alpha^2q^2}} \approx -i$.

Thus $\mu_{n,Q}(t) \to n!\quad r^{-n}C_n^{-(r^2t/2)}(1) = (-1)^n r^{-n}(r^2t)^{(n)}$ by the scalings $t = r^2t'$, $v = -r^{-1}v'$ for the exponential process. This checks with

$$(1-2y+y^2)^{r^2t/2} = (1-y)^{r^2t} = \Sigma \frac{(-1)^n}{n!}(r^2t)^{(n)}y^n.$$

4. The canonical variable $x \operatorname{csh}^2 \frac{Qz}{2} \to x(1+\frac{z}{r})^2$.

   We have the convergence of the canonical densities, and convergence of the kernels, but we will not explicitly discuss these for this case.

BERNOULLI → SYMMETRIC BERNOULLI

In the limit $\alpha \to 0$, $\operatorname{csh} \to \cosh$, $Q \to \sqrt{2b}$, $b \equiv -\beta$.

The generator $-\frac{\alpha}{\beta}z - \frac{2}{\beta}\log \operatorname{csh}\frac{Qz}{2} \to \frac{2}{b}\log \cosh\sqrt{b/2}\, z$. The probabilities $p, \bar{p}$ converge to $\frac{1}{2}$. The polynomials are

$$(\tfrac{b}{2})^{n/2}K_n(\sqrt{\tfrac{2}{b}}\, x, \tfrac{2}{b}\, t) = (\tfrac{b}{2})^{n/2}\, \Sigma \binom{n}{k}(-1)^{n-k}(\tfrac{t}{b}+\tfrac{x}{\sqrt{2b}})^{(k)}(\tfrac{t}{b}-\tfrac{x}{\sqrt{2b}})^{(n-k)}.$$

This limit is simply a balancing of the general process, so the formulas are basically the same.

BERNOULLI → POISSON

1. The limit $\beta \to 0$ is a "singular" limit, from the point of view that π changes from quadratic to linear. The generator

$$-\frac{\alpha}{\beta}z - \frac{2}{\beta}\log[p(e^{Qz}-1)+1] + \frac{Q}{\beta}z = rz - \frac{2}{\beta}\log[p(e^{Qz}-1)+1]$$

   As $\beta \to 0$, $Q \to \alpha$. $r = \frac{-\alpha+(\alpha^2-2\beta)^{\frac{1}{2}}}{\beta} \approx -\frac{1}{\alpha}$. $s = \frac{-\alpha-(\alpha^2-2\beta)^{\frac{1}{2}}}{\beta} \approx \pm\infty$.

   $\alpha = \alpha q \to 1$. Thus,

$$L \approx -\frac{z}{\alpha} + \frac{2}{\beta}\frac{\beta r}{2}q(e^{Qz}-1) - \frac{2}{\beta}\frac{q^2\beta^2r^2}{4} \dots \to \alpha^{-2}(e^{\alpha z}-1-\alpha z).$$

$$V = \frac{s\partial_Q^+}{s+\partial_Q^+} = \frac{\partial_Q^+}{1+s^{-1}\partial_Q^+} \to \partial_\alpha^+.$$

Proposition 31: For any Q and corresponding process $x_Q(t)$,

$$\tfrac{1}{2}x_Q(-\frac{2t}{\beta q^2}) - \frac{t}{\beta q^2} \underset{\beta\to 0}{\to} N(t), \text{ a standard Poisson process.}$$

Proof:

$$\tfrac{1}{t}\log\langle\exp[\tfrac{z}{2}x_Q(\tfrac{-2t}{\beta q^2}) - \tfrac{zt}{\beta q^2}]\rangle = \frac{2}{\beta q^2}\log(pe^{z/2}+\bar{p}e^{-z/2}) - \frac{z}{\beta q^2}$$

$$= -\frac{2}{\beta q^2}\log[p(e^z-1)+1] \approx -\frac{2\beta qr}{2\beta q^2}(e^z-1) \to e^z-1,$$

as above.

This may be written as a "large-time" limit. Set $q = 1$, $t = \frac{2t'}{\beta}$.

<u>Corollary</u>: $\frac{1}{2}(x_1(t)+t) \underset{t\to\infty}{\to} N(t)$ providing p depends on t so that $\underset{t\to\infty}{\ell t}\ tp(t) = 1$.

<u>Proof</u>: $\log\langle e^{\frac{z}{2}x_1(t)} e^{\frac{z}{2}t}\rangle = t\log[p(e^z-1)+1] \approx tp(e^z-1)+tp^2 \ldots$

Since $tp \to 1$, tp^2 and higher powers $\to 0$.

<u>Remark</u>: This is the usual "Poisson limit theorem" for Bernoulli variables in the context of processes.

2. The J_n's convergence as $\beta \to 0$ is as follows.

<u>Proposition 32</u>: $J_n(x,t)$ for the general w-process converges as $\beta \to 0$ to

$$\alpha^n P_n(\alpha^{-1}x+\alpha^{-2}t, \alpha^{-2}t),$$

where P_n are the Poisson-Charlier polynomials.

<u>Proof</u>: The scaling $t = \alpha^{-2}t'$, $x-t = \alpha^{-1}x'$ in terms of the processes is $N(\alpha^{-2}t) = \alpha^{-2}t + \alpha^{-1}w(t)$. Hence $P_n(\alpha^{-1}x+\alpha^{-2}t, \alpha^{-2}t)$ and the factor α^n comes from the scaling $v = \alpha v'$.

3. We check convergence of the M's.

$$M(v) = q\log[(1-\frac{v}{r})^r(1-\frac{v}{s})^{-s}] \approx \alpha^{-1}\log[(1+\alpha v)^{-\alpha^{-1}}e^v] = \alpha^{-1}v-\alpha^{-2}\log(1+\alpha v),$$

since as $\beta \to 0$, $s \to \infty$. The moments in this context are most conveniently expressed by $\mu_n(t) = J_n(0,-t)$ so convergence of the moments follows from the general convergence of the J_n's. We note that because of the added drift, $\mu_n(t)$ for the mean-zero Poisson case equals $J_n(0,-t) = \alpha^n P_n(-\alpha^{-2}t, -\alpha^{-2}t)$; compare with the expansion for the gamma transform (Chapter V).

<u>Remark</u>: We could define a Poisson transform just as we did for the gamma transform,

$$\langle g[N(t)]\rangle = \Sigma \frac{P_n(-t,-t)}{n!} \partial_1^{+n} g(t), \quad P_n(-t,-t) = \Sigma \binom{n}{k} t^{n-k}(-t)^{(k)}.$$

Another expression for $\langle g[N(t)]\rangle$ is of interest.

$$\langle g[N(t)]\rangle = e^{-t}\sum_0^\infty \frac{t^k}{k!} g(k) = e^{-t}g(tT)e^t, \quad T = \frac{d}{dt}.$$

Of course, tT is the "A" operator in the t-variable. We have, e.g.

$$\langle N^k\rangle = e^{-t}(tT)^k e^t, \quad \langle N^{(k)}\rangle = t^k = e^{-t}(tT)^{(k)}e^t.$$

We now have the cases of the Central Limit Theorem in our context.

EXPONENTIAL → BROWNIAN MOTION

1. The generator $L = rz - r^2\log(1+\frac{z}{r})$ converges as $r \to \infty$:

$$rz - r^2\log(1+\frac{z}{r}) \approx rz - r^2\,\frac{z}{r} + r^2\,\frac{z^2}{2r^2} = \frac{z^2}{2} \quad \text{as expected.}$$

$V = \frac{rz}{r+z} \to z$. So we have

Proposition 33: Let $e(t)$ be a standard exponential process. Then

(1) $-r^{-1}[e(r^2t)-r^2t] \to b(t)$ as $r \to \infty$, where $b(t)$ is Brownian motion.

(2) $s^{-\frac{1}{2}}[e(st)-st] \to b(t)$ as $s \to \infty$.

(3) $t^{-\frac{1}{2}}[e(t)-t] \to b(1)$ as $t \to \infty$.

Proof: (1) Follows since $w(t) = -r^{-1}[e(r^2t)-r^2t]$ is the "full" exponential process.

(2) Set $r^2 = s$. Then we have $s^{-\frac{1}{2}}[st-e(st)] \to b(t)$. Use the symmetry of Brownian motion to change to $e(st)-st$.

(3) Put $t = s'$. $1 = t'$ in (2).

2. As we saw in Proposition 30, the J_n's for the exponential process are $(-r)^{-n}L_n(r^2t-rx,r^2t)$. We thus have

Proposition 34: For L_n, the Laguerre polynomials,

$(-r)^{-n}L_n(r^2t-rx,r^2t) \to H_n(x,t)$ as $r \to \infty$, where H_n are the Hermite polynomials.

Example: We use this as an example of (exact!) perturbation theory. Put $r^{-1} = \varepsilon$. We have the number operator in the exponential case $x\frac{\partial}{\partial x} + (\varepsilon x-t)\frac{\partial^2}{\partial x^2}$. We write $\varphi_n = (-r)^{-n}L_n(r^2t-rx,r^2t)$. So

$$[x\frac{\partial}{\partial x} + (\varepsilon x-t)\frac{\partial^2}{\partial x^2}]\varphi_n = n\varphi_n(x,t).$$

Set $\varphi_n = H_n(x,t) + \sum_1^\infty \varepsilon^j F_{nj}$.

Apply the operator $xD-tD^2-n$ to get:

$$(xD-tD^2-n)\varphi_n = 0 + \sum_1^\infty \varepsilon^j (xD-tD^2-n)F_{nj} = -\varepsilon xD^2\varphi_n.$$

Equating powers of ε yields, dropping the "n" subscript on the F's,

$$\begin{cases} (xD-tD^2-n)F_{j+1} = -xD^2F_j, & j \geq 0 \\ F_o = H_n(x,t). \end{cases}$$

Put $\bar{A} = xD-tD^2$. So we want to compute the resolvents $(n-\bar{A})^{-1}$ to solve for $F_{j+1} = (n-\bar{A})^{-1}xD^2F_j$. A recursive procedure is the following:

(1) From $xH_n-tnH_{n-1} = H_{n+1}$ we have

$$xn^{(k)}H_{n-k} = tn^{(k+1)}H_{n-k-1} + n^{(k)}H_{n-k+1}.$$

(2) Since $\bar{A}$ is the number operator for the H_n's,

$$(n-\bar{A})^{-1}H_{n-k} = k^{-1}H_{n-k}.$$

(3) We thus have

$$F_1 = (n-\bar{A})^{-1}xD^2H_n = (n-\bar{A})^{-1}xn(n-1)H_{n-2} = \frac{t}{3}n^{(3)}H_{n-3} + n^{(2)}H_{n-1}$$

$$F_2 = \frac{t^2}{18}n^{(6)}H_{n-6} + t\left(\frac{n^{(5)}}{12} + \frac{n-1}{4}n^{(4)}\right)H_{n-4} + \frac{n-1}{3}n^{(3)}H_{n-2}$$

and so on.

E.g. $n = 2$, $F_{20} = x^2-t$, $F_{21} = 2H_1 = 2x$.

$$\varepsilon^2L_2(\varepsilon^{-2}t-\varepsilon^{-1}x,\varepsilon^{-2}t) = x^2-t+2\varepsilon x, \quad \text{as computed previously.}$$

(4) The expansion is exact. We have

$\varphi_n = (-\varepsilon)^nL_n(\varepsilon^{-2}t-\varepsilon^{-1}x,\varepsilon^{-2}t) \to H_n(x,t)$ as $\varepsilon \to 0$ and so must be analytic in ε near 0 since L_n is a polynomial function in both its arguments.

3. The moments of the full M(v), $J_n(0,-t) = (-r)^{-n}L_n(-r^2t,-r^2t) = (-r)^{-n}e_n(r^2t)$ (see the section on the gamma transform in Chapter V), converge to $\mu_{2k}(t) = \frac{2k!}{2^kk!}$, $\mu_{2k+1}(t) = 0$ as $r \to \infty$. We thus see, since $\deg(e_{2k}) = k = \deg(e_{2k+1})$, that the leading term of $e_{2k}(t)$ is $t^k\frac{2k!}{2^kk!}$.

4. In this case we can discuss convergence of the kernels k_t and K_t. For e(t), $k_t = e^{-y}\psi_t(xy)$, where $\psi_t(x) = \Gamma(t)x^{(1-t)/2}I_{t-1}(2\sqrt{x})$. For the full process we have $k_t(x,y) = e^{-y}\psi_{r^2t}[y(r^2t-rx)]$ which thus converges to $\exp(\frac{xy}{t} - \frac{y^2}{2t})$ as

$r \to \infty$. Similarly for the kernel K_t, replace t by r^2t, x,y by r^2t-rx, r^2t-ry, respectively.

SYMMETRIC BERNOULLI → BROWNIAN MOTION

1. We have $L = \frac{2}{b}\log\cosh\sqrt{b/2}\,z = \frac{z^2}{2} + \frac{2}{b}(\frac{b}{2})^{3/2}\ldots \approx \frac{z^2}{2}$ as $b \to 0$.
   This is the original Central Limit Theorem where scaled jumps ± 1 are converging to a Gaussian distribution. We have

Proposition 35: Let $d(t)$ be a standard symmetric Bernoulli process.

(1) $\sqrt{b/2}\,d(\frac{2t}{b}) \to b(t)$, Brownian motion, as $b \to 0$.

(2) $s^{-\frac{1}{2}}d(st) \to b(t)$ as $s \to \infty$.

(3) $t^{-\frac{1}{2}}d(t) \to b(1)$ as $t \to \infty$.

Proof: (1) follows from the symmetric Bernoulli scaling.

(2) follows by setting $s = \frac{2}{b}$. Then $s^{-\frac{1}{2}}d(st) \to b(t)$ as $s \to \infty$.

(3) follows by $t = s'$, $1 = t'$ in (2).

2. The K_n's converge to H_n. We have

Proposition 36: For K_n the Krawtchouk polynomials,

$$(\tfrac{b}{2})^{n/2}K_n(\sqrt{2/b}\,x, \tfrac{2}{b}t) \to H_n(x,t), \text{ the Hermite polynomials,}$$

as $b \to 0$.

We could set $\frac{b}{2} = \varepsilon^2$ to have $\varepsilon^n K_n(\varepsilon^{-1}x, \varepsilon^{-2}t) \to H_n(x,t)$ as $\varepsilon \to 0$.

3. The moments $\frac{(2k)!}{k!}(\frac{b}{2})^k(\frac{t}{b})_{(k)} \to \frac{2k!}{2^k k!}t^k$, appropriately, as $b \to 0$.

POISSON → BROWNIAN MOTION

1. We have $L = \alpha^{-2}(e^{\alpha z}-1-\alpha z) \to \frac{z^2}{2}$ as $\alpha \to 0$. This is interesting to consider from the possible physical existence of an "elementary length," i.e. the possibility that space is "quantized."
   The process $\alpha[N(\alpha^{-2}t)-\alpha^{-2}t]$ is a process which steadily drifts along while being frequently (depending on the smallness of α) kicked by Poisson "particles." This seems quite like one might imagine Brownian motion to "really" be. We have

<u>Proposition 37</u>: Let $N(t)$ be a standard Poisson process. Then

(1) $\alpha[N(\alpha^{-2}t)-\alpha^{-2}t] \to b(t)$, Brownian motion, as $\alpha \to 0$.

(2) $s^{-\frac{1}{2}}[N(st)-st] \to b(t)$, as $s \to \infty$.

(3) $t^{-\frac{1}{2}}[N(t)-t] \to b(1)$, as $t \to \infty$.

<u>Proof</u>: Set $s = \alpha^{-2}$ and (1)-(3) follow as for the previous cases.

2. We have

<u>Proposition 38</u>: For P_n the Poisson-Charlier polynomials,

$$\alpha^n P_n(\alpha^{-1}x+\alpha^{-2}t, \alpha^{-2}t) \to H_n(x,t), \text{ Hermite polynomials, as}$$

$\alpha \to 0$.

FURTHER REMARKS

1. If the parameters are restricted so that the measures $p_t(dx)$ corresponding to the generators $L(z)$ are probability measures, then by techniques such as using martingales to determine processes, one can deduce weak convergence of the processes from convergence of the generators. It is part of our motivation here to suggest that similar properties of convergence of "reasonable" measures, not necessarily positive, should hold.

2. In the various limits we have considered the Lagrangians $R(\alpha) = \alpha U(\alpha)-M(\alpha)$ converge to the corresponding Lagrangians. Since R determines L and vice versa, up to centering constants, we can think then that the dependence of energy upon the speed really is characteristic for the entire process in the sense that all observable functionals of the processes, i.e. expected values, converge if and only if the energy functionals converge. This is the "physical" significance of weak convergence of the processes.

CHAPTER VII. DISCRETE THEORY. MARTINGALES. STOCHASTIC INTEGRALS.

As we have seen, Brownian motion is very special; in our context it is because the moment and orthogonal polynomials coincide. We now observe the following facts about Brownian motion $b(t)$ and seek to understand them in terms of our formulation of more general processes. Namely we have

(1) Define $I_{n+1}(t) = \int_0^t I_n db$, $I_o \equiv 1$, $I_1(t) = b(t)$. Then

$n!I_n(t) = H_n[b(t),t]$; Hermite polynomials.

(2) For martingales $F[b(t)]$, $dF = F'[b(t)]db$.

The I_n's in (1) are the iterated stochastic integrals of the process db. We will see in the following that the I_n's have the properties that

(a) they are martingales,

(b) they are orthogonal functionals, i.e. $\langle I_n I_m \rangle = j_n \delta_{nm}$.

We thus expect that the J_n's of Chapter IV would be (n! times) the iterated integrals of our general w(t)-process. However, we will also see that regardless of the value of β,

$$\langle I_n^2 \rangle = \langle \frac{J_n^2}{(n!)^2} \rangle = \frac{t^n}{n!} \quad \text{whereas} \quad \langle \frac{J_n^2}{(n!)^2} \rangle = \frac{1}{n!}t(t+\frac{\beta}{2})\dots(t+\frac{n-1}{2}\beta),$$

which agrees only for $\beta = 0$. It can be seen that the J_n's we have are iterated stochastic integrals, but we need more clarification as to the exact relationship between our J_n's and the iterated integrals as in (1) above.

We proceed to study the integrals by approximating by discrete processes. This approach will illustrate precisely what is happening. As a bonus we will deduce some new limit theorems for symmetric functionals of processes.

ITERATED INTEGRATION AND EXPONENTIALS

Let's define inductively $I_o \equiv 1$, $I_1 = x$,

$$I_{n+1} = \int_0^x I_n(y)dy.$$

Clearly then,

$$I_n = \frac{x^n}{n!} \quad \text{and} \quad E = \Sigma v^n I_n = e^{vx}.$$

Since the I_n's are defined iteratively we have

$$dE = vE\,dx$$

which is often used as a defining property of the exponential.

Given a measure df we can define similarly

$$I_n(x) = \int_0^x I_{n-1}(y)df(y), \quad I_o = 1, \quad I_1 = f(x).$$

Here $E = \Sigma v^n I_n$ and we have

$$dE = vE\,df, \quad E = e^{vf(x)}, \quad I_n = \frac{f(x)^n}{n!}.$$

The main observations are that iterated integrals correspond to algebraic powers and that the generating function of a sequence of iterated integrals is our "exponential" function.

DISCRETE ITERATED INTEGRALS

We now consider an arbitrary sequence of numbers x_j or a triangular array x_{jn}, $1 \leq j \leq n$. The x_j's are the differentials "dx" or "df." We define

$$I_o \equiv 1. \quad I_1(n) = \sum_1^n x_j. \quad I_k(n) = \sum^n I_{k-1}(j-1)x_j.$$

We set $I_k(n) = 0$ for $k > n$. So the sum for $I_k(n)$ starts with $j = k$. We will see later why the index is j-1 for I_{k-1}; these are called "non-anticipating" integrals for this reason.

<u>Proposition 39</u>: For $E_n = \sum_o^\infty v^k I_k(n)$,

(1) $E_n = \prod_1^n (1+vx_j)$,

(2) $\delta E_n = E_n - E_{n-1} = vE_{n-1}x_n$.

<u>Proof</u>:

$$E_o = 1. \quad \delta E_n = E_n - E_{n-1} = \Sigma\, v^k[I_k(n) - I_k(n-1)] = \Sigma\, v^k I_{k-1}(n-1)x_n$$
$$= vE_{n-1}x_n.$$

That is, $E_n = (1+vx_n)E_{n-1}$ and (1) and (2) follow.

This is thus our discrete exponential. Expanding, we conclude

<u>Corollary</u>: $I_k(n) = k^{th}$ elementary symmetric polynomial in the variables $(x_1,\ldots,x_n)$.

E.g. Newton's theorem on symmetric polynomials may be restated in the form:

Any symmetric polynomial can be expressed as a sum

$$\Sigma\, a_{k_1\ldots k_n} I_1^{k_1} I_2^{k_2} \cdots I_n^{k_n}$$

of products of powers of discrete iterated integrals. It is this formulation which extends to (symmetric) functionals of processes yielding Wiener's homogeneous chaoses.

MARTINGALES AND MARTINGALE DIFFERENCES

In the case where the x_j are martingale differences we have some interesting properties of the I_k's and E_n's. Recall that $Y_n = \sum_1^n x_j$ is a <u>martingale</u> if

(1) we have a fixed probability space $(\Omega, \mathcal{F}, P)$,

(2) we have an increasing family of σ-fields $\mathcal{F}_1 \subset \mathcal{F}_2 \ldots \subset \mathcal{F}$ such that $\sigma\{Y_n\}$ = σ-field generated by $Y_n \subset \mathcal{F}_n$, $n \geq 1$,

(3) $E(Y_n | \mathcal{F}_m) = Y_{\min(n,m)}$,

(4) the x_j are <u>martingale differences</u> and satisfy

(a) $E(x_j | \mathcal{F}_k) = 0$ for $k < j$,

(b) $\langle x_j \rangle = 0$.

For each n, assume that our x_{jn}, $1 \leq j \leq n$ are a sequence of martingale differences. We have

<u>Proposition 40</u>: Given a sequence x_j and an array x_{jn} of martingale differences,

(1) Each $E_n(k) = \prod_1^k (1+vx_{jn})$ is a martingale (in k),

(2) Each $I_j(k)$ is a martingale (in k),

(3) If $\sum_1^\infty |x_j|$ converges, then $E_n = \prod_1^n (1+vx_j)$ converges as $n \to \infty$.

(4) For $E_n = \prod_1^n (1+vx_j)$, if $\sup_n \langle |E_n| \rangle < \infty$, then E_n converges a.s.

<u>Proof</u>: (1) $E(E_k | \mathcal{F}_{k-1}) = \prod_1^{k-1} (1+vx_{jn})$. $E(1+vx_{kn} | \mathcal{F}_{k-1}) = E_{k-1}$.

(2) $E(I_j(k) | \mathcal{F}_{k-1}) = E(\sum_1^k I_{j-1}(\ell-1)x_\ell | \mathcal{F}_{k-1}) = \sum_1^{k-1} I_{j-1}(\ell-1)x_\ell = I_j(k-1)$, since $\sigma\{I_j(\ell)\} \subset \mathcal{F}_\ell$.

(3) We have $|E_n - E_m| = |\prod_{m+1}^n (1+vx_j) - 1| |E_m|$.

$$|E_m| \le \prod_1^\infty (1+|v|\,|x_j|) \le e^{|v|\sum_1^\infty |x_j|} = M.$$

And $\left|\prod_{m+1}^{n} (1+vx_j)-1\right| \le M \sum_{m+1}^{n} |vx_j| \to 0$ as $m,n \to \infty$,

where we use

$$\left|\prod_1^N a_k - \prod_1^N b_k\right| = \left|\Sigma\, b_1 \ldots b_{k-1}(a_k-b_k)a_{k+1}\ldots a_N\right| \le M\, \Sigma |a_k-b_k|$$

with $b_j = 1+vx_j$, $a_j \equiv 1$, so that every product $|b_1 \ldots b_{k-1}| \le M$.

(4) follows from the martingale convergence theorem.

Remark: (2) explains why non-anticipating integrals are used, namely, to preserve the martingale property. Note that (3) is independent of the martingale condition.

PROCESSES WITH INDEPENDENT INCREMENTS

We now assume the stronger condition that x_j are independent, $\langle x_j \rangle = 0$. Then $\sum_1^n x_j$ is a process with independent increments.

Proposition 41: When x_j's are independent, $\langle x_j \rangle = 0$, var $x_j = \sigma_j^2$, then

(1) $I_k(n)$ are orthogonal, i.e. for fixed n, $\langle I_k(n) I_\ell(n) \rangle = 0$, $k \neq \ell$.

(2) $\langle I_k(n)^2 \rangle = k^{th}$ elementary symmetric polynomial in the variables $\sigma_1^2, \ldots, \sigma_n^2$.

Proof:

$$\langle E_n^2 \rangle = \Sigma\Sigma\, v^{k+\ell} \langle I_k(n) I_\ell(n) \rangle = \langle \prod_1^n (1+2vx_j+v^2x_j^2) \rangle$$

$$= \prod_1^n (1+v^2\sigma_j^2) = \Sigma\, v^{2k} \langle I_k^2 \rangle.$$

Remark: So if the x_j are identically distributed, var $x_j = \sigma^2$, we have $\langle I_k^2(n) \rangle = \binom{n}{k} \sigma^{2k}$.

We now proceed to study some examples in the deterministic and probabilistic contexts and we'll see that these integrals are of quite general occurrence.

DETERMINISTIC EXAMPLES

1. The simplest of all cases is to put x_j identically 1. Then we have $E_n = (1+v)^n$, $I_k(n) = \binom{n}{k}$. The iteration relation is thus

$$I_k(n) = \binom{n}{k} = \sum_{j=k}^{n} \binom{j-1}{k-1},$$

which is the same as Pascal's

$$\binom{n}{k} = \binom{n-1}{k-1} + \binom{n-1}{k}.$$

2. The next simplest case is to set $x_{jn} = \frac{x}{n}$. Then $E_n = (1+\frac{vx}{n})^n$ and $I_k(n) = \binom{n}{k}(\frac{x}{n})^k$. In this case, as $n \to \infty$, we have

$$E_n \to e^{vx}, \quad I_k(n) \to \frac{x^k}{k!} = k^{th} \text{ iterated } \int \text{ of } dx,$$

so we could say that $\frac{x}{n} \to dx$ as $n \to \infty$.

3. An example of importance in number theory and theta functions arises from the choice $x_j = q^{j-1}$. $I_o = 1$, $I_1 = \sum_1^n q^{j-1} = \frac{q^n-1}{q-1}$,

$$I_2 = \sum^n \frac{q^{k-1}-1}{q-1} q^{k-1} = \frac{1}{q-1} \sum (q^2)^{k-1} - q^{k-1} = \frac{(q^n-1)(q^n-q)}{(q^2-1)(q-1)}.$$

We will check that $I_k(n) = \frac{(q^n-1)(q^n-q)\ldots(q^n-q^{k-1})}{(q^k-1)(q^{k-1}-1)\ldots(q-1)}$.

<u>Proposition 42</u>: For $x_j = q^{j-1}$,

$$E_n = \prod_1^n (1+vq^{j-1}) = \sum \binom{n}{k}_q q^{\binom{k}{2}} v^k$$

where $\binom{n}{k}_q = \frac{(q^n-1)(q^n-q)\ldots(q^n-q^{k-1})}{(q^k-1)(q^k-q)\ldots(q^k-q^{k-1})}$.

<u>Proof</u>: Observe first that $\binom{n}{k}_q q q^2 \ldots q^{k-1} = I_k(n)$ as given above.

Now we check that $\sum v^k I_k(n)$ satisfies $\delta E_n = vE_{n-1}q^{n-1}$.

We have

$$E_n - E_{n-1} = \sum \frac{(q^n-1)\ldots(q^n-q^{k-1})}{(q^k-1)\ldots(q-1)} v^k - \frac{(q^{n-1}-1)\ldots(q^{n-1}-q^{k-1})}{(q^k-1)\ldots(q-1)} v^k$$

$$= \Sigma \frac{q^{k-1}(q^n-1)}{q^k-1} \frac{(q^{n-1}-1)\dots(q^{n-1}-q^{k-2})}{(q^{k-1}-1)\dots(q-1)} v^k$$

$$- \frac{(q^{n-1}-1)\dots(q^{n-1}-q^{k-2})}{(q^{k-1}-1)\dots(q-1)} \frac{q^{n-1}-q^{k-1}}{q^k-1} v^k$$

$$= v\, \Sigma\, I_{k-1}(n-1)v^{k-1} \frac{q^{n+k-1}-q^{n-1}}{q^k-1} = vq^{n-1}E_{n-1}.$$

Remarks:

1. The coefficient $\binom{n}{k}_q$ is Gauss' binomial coefficient. As $q \to 1$ we have $E_n \to (1+v)^n$ and so $\binom{n}{k}_q \to \binom{n}{k}$.

2. The exponential function $E_n = \prod_1^n (1+vq^{j-1})$, for $0 < q < 1$, converges as $n \to \infty$ to the theta product $\prod_0^\infty (1+vq^j)$. The limits of $\binom{n}{k}_q q^{\binom{k}{2}}$ are thus the iterated sums of the variables q^{j-1}; we have, for $0 < q < 1$,

$$\lim_{n\to\infty} I_k(n) = q^{\binom{k}{2}} \prod_{j=1}^{k} (1-q^j)^{-1} = \sum_{\substack{j_1,\dots j_k \geq 0 \\ j_\ell \text{ distinct}}} q^{j_1}q^{j_2}\dots q^{j_k}$$

and $\lim_{n\to\infty} \binom{n}{k}_q = \prod_{j=1}^{k} (1-q^j)^{-1}$.

3. We can choose $x_{jn} = q_n^{j-1}(t^{1/n}-1)$ with $q_n = t^{1/n}$. Then we have

$$I_k(n) = \frac{(t-1)(t-t^{1/n})\dots(t-t^{(k-1)/n})}{(t^{k/n}-1)\dots(t^{1/n}-1)}(t^{1/n}-1)^k$$

and $E_n = \prod_1^n [1+v(t^{j/n}-t^{(j-1)/n})]$.

We have chosen x_{jn} to be the differences $t^{j/n}-t^{(j-1)/n}$ of a partition of $[1,t]$. We thus expect

$$I_k(n) \to \frac{(t-1)^k}{k!} \quad \text{and} \quad E_n \to e^{vt-v} \quad \text{as } n \to \infty.$$

The first limit follows from the expression for $I_k(n)$; we have to note that

$$\lim_{n\to\infty} \frac{t^{1/n}-1}{t^{k/n}-1} = \lim_{x\to 0} \frac{t^x-1}{t^{kx}-1} = \lim_{x\to 0} \frac{\log t\cdot t^x}{k\log t\cdot t^x} = \frac{1}{k}.$$

To prove convergence of E_n we use the following.

Proposition 43: If $\sup_n \sum_{j=1}^{n} |x_{jn}| < \infty$, then $I_k(n)$ converge to I_k if and only if $E_n(v)$ converges to an entire function $E(v) = \Sigma\, v^k I_k$ uniformly on compact subsets of $\{v\} \subset \mathbb{C}$.

Proof: (1) Assume $I_k(n)$ converge to I_k. We have

$$|E_n(v)| \leq \prod_1^n (1+|v||x_{jn}|) \leq e^{|v|\Sigma|x_{jn}|}$$

which is uniformly bounded for $|v| \leq K$. Since $E_n(v)$ are holomorphic, being polynomials, they converge normally (i.e., uniformly on compact subsets of $\mathbb{C}$) to a holomorphic function $E(v)$. By Cauchy's integral formula,

$$I_k(n) = \frac{1}{2\pi i}\int_o \frac{E_n(v)}{v^{n+1}}\,dv \quad \text{converge to} \quad \frac{1}{2\pi i}\int_o \frac{E(v)}{v^{n+1}}\,dv.$$

Since $I_k(n) \to I_k$, we have $E(v) = \Sigma\, v^k I_k$. (This shows uniqueness of the limit, so we have $E_n \to E$ not just a subsequence.)

(2) follows as in the last part above.

We apply this to the above example with $x_{jn} = t^{j/n} - t^{(j-1)/n}$ so that $\sum_1^n |x_{jn}| = t-1$. We thus conclude $E_n(v) \to e^{vt-v}$, uniformly for $|v| \leq K$.

FURTHER REMARKS. BASIC OPERATORS

We observe that if we assume x_j to satisfy the relations $[x_j, x_k] = 0$, $x_j^2 = 0$ then we have, for $X_n = \sum_1^n x_j$,

$$I_k(n) = X_n^{\,k} \quad \text{and} \quad E_n(v) = e^{vX_n}.$$

This type of algebra is not unreasonable in our context. If we think of the x_j's as being infinitesimals "dx" then the relation $x_j^{\,2} = 0$ is the usual "neglecting infinitesimals of 2nd order." However, we are keeping products $x_j x_k$, so this is not exactly consistent. It is actually a "Pauli exclusion principle."

If we consider the expansion $E = \Sigma \frac{v^k J_k}{k!}$, $J_o = 1$, we see that $\frac{\partial E}{\partial v} = \Sigma \frac{v^k J_{k+1}}{k!}$ and $vE = \Sigma \frac{v^k k J_{k-1}}{k!}$ so the Heisenberg duals ∂_v and v act in "opposite" fashion on the J's sending $J_k \to J_{k+1}$, $J_k \to kJ_{k-1}$ respectively. We have

<u>Definition</u>: (1) $J_k(n) = k!I_k(n)$.

(2) Define the operators d by $dJ_k = kJ_{k-1}$ and c by $cJ_k = J_{k+1}$.

<u>Proposition 44</u>: 1. $\sum_1^n \frac{x_j}{1+vx_j} \cdot E = \sum \frac{v^k}{k!} cJ_k$

2. $\frac{1}{n}\sum_1^n (1+vx_j)\frac{\partial}{\partial x_j}E = \frac{1}{n}(\mathrm{div} + v\underset{\sim}{x}\cdot\underset{\sim}{D})E = \sum \frac{v^k}{k!} d\,J_k$

3. $\underset{\sim}{x}\cdot\underset{\sim}{D} = \sum x_j \frac{\partial}{\partial x_j}$ is the number operator, i.e. $\underset{\sim}{x}\cdot\underset{\sim}{D}J_k(n) = kJ_k(n)$.

4. We have the recursion formula $J_{k+1} = \sum_{p=0}^{k} (\sum_{j=1}^{n} x_j^{p+1})(-1)^p d^p J_k$,

i.e., $c = \sum_p (\sum_{j=1}^{n} x_j^{p+1})(-d)^p$

5. $\mathrm{div}\, J_k = (n-\underset{\sim}{x}\cdot\underset{\sim}{D})dJ_k$

6. $d = (n-\underset{\sim}{x}\cdot\underset{\sim}{D})^{-1}\mathrm{div}$

<u>Proof</u>: 1. $\frac{\partial E}{\partial v} = \frac{\partial}{\partial v}\Pi(1+vx_j) = \sum_{j=1}^{n} \frac{x_j}{1+vx_j}E = \sum \frac{v^k}{k!} J_{k+1}$

2. $\frac{\partial E}{\partial x_j} = \frac{v}{1+vx_j}E$ so $vE = (1+vx_j)\frac{\partial E}{\partial x_j} = \frac{1}{n}\sum (1+vx_j)\frac{\partial E}{\partial x_j}$.

3. $J_k(n)$ is homogeneous of degree k so we have, for $D_j = \frac{d}{dx_j}$,

$$\lambda^{\underset{\sim}{x}\cdot\underset{\sim}{D}}J_k(n) = \lambda^{x_1D_1}\lambda^{x_2D_2}\cdots\lambda^{x_nD_n}J_k(n) = k!\,\sum(\lambda x_{j_1})\cdots(\lambda x_{j_k}) = \lambda^k J_k(n).$$

Differentiating with respect to λ at $\lambda = 1$ yields $\underset{\sim}{x}\cdot\underset{\sim}{D}J_k = kJ_k$.

4. $\frac{\partial E}{\partial v} = \sum \frac{x_j}{1+vx_j}E = \sum_j \sum_p (-1)^p v^p x_j^{p+1} \sum_k \frac{v^k}{k!} J_k$

$$= \sum \frac{v^m}{m!} \sum_{p\le m} (-1)^p \sum_j x_j^{p+1} m^{(p)} J_{m-p} = \sum \frac{v^m}{m!} J_{m+1}.$$

Now we note that $m^{(p)}J_{m-p} = d^p J_m$.

5. $vE = \sum \frac{v^k}{k!} dJ_k = \frac{1}{n}\sum \frac{v^k}{k!}\mathrm{div}\, J_k + \frac{1}{n}\sum \frac{v^k}{k!}\underset{\sim}{x}\cdot\underset{\sim}{D}\, dJ_k$ by (2).

Thus, $\mathrm{div}\, J_k = (n-\underset{\sim}{x}\cdot\underset{\sim}{D})d\, J_k$.

6. follows from (5).

Remarks:

1. The c and d here play the role of C and D of moment theory, $\overline{C}W,V$ of the orthogonal theory. Under proper convergence we would recover these operators as limiting cases of c,d.

2. The expression of c in terms of d can be written

$$c = X + \sum_{p\geq 1} X(p+1)(-d)^p \quad \text{with} \quad X = \sum_1^n x_j, \quad X(m) = \sum_1^n x_j^m.$$

This is reminiscent of, and the prototype of, the expression $C = x + tL'$.

We consider now E as a function of the variable x_n, as we would for the exponential $e^{zw(t)-tL(z)}$, replacing $w(t)$ by x. We have as a discrete version of $\frac{\partial u}{\partial t} = Lu$,

Proposition 45:

1. $E = E_n$ satisfies $\delta E = x_n \frac{\partial}{\partial x_n} E$, i.e. the generator is $x \frac{\partial}{\partial x}$.

2. $\delta J_k(n) = x_n \frac{\partial}{\partial x_n} J_k(n)$

3. $\frac{\partial}{\partial x_n} J_k(n) = kJ_{k-1}(n-1)$.

Proof:

1. $\delta E = vE_{n-1}x_n$ by construction. We observe that

$$vE_{n-1} = \frac{v}{1+vx_n} E_n = \frac{\partial}{\partial x_n} E_n.$$

2. follows by applying δ to $E = \sum \frac{v^k}{k!} J_k$.

3. By construction $\delta I_k(n) = I_{k-1}(n-1)x_n$ so that

$$\delta J_k(n) = kJ_{k-1}(n-1)x_n = x_n \frac{\partial}{\partial x_n} J_k(n) \quad \text{by (2).}$$

So our coharmonic functions are those satisfying $\delta u = x_n \frac{\partial}{\partial x_n} u$.

PROBABILISTIC THEORY

There are two aspects to the probabilistic theory. The first topic we will consider is deriving the iterated integrals for a given process with stationary independent increments. The second topic is limit theorems, deriving iterated integrals of one process from those of simpler processes.

We proceed with the first aspect. Assume we have $w(t)$ with generator $L(z)$. Then the increments $x_{jn} = w(\frac{j}{n} t) - w(\frac{j-1}{n} t)$ are independent, identically distributed and satisfy

$$\langle e^{zx_{jn}}\rangle = e^{\frac{t}{n}L(z)} .$$

If we further center $w(t)$ so that $\langle w(t)\rangle = 0$ then $w(t)$ is a martingale and x_{jn} are martingale differences. By Proposition 40, each $I_k(\ell)$ and $E_n(\ell)$ are martingales (as ℓ varies). Proposition 41 implies the orthogonality of the $I_k(n)$ and

$$\text{var } I_k(n) = \binom{n}{k}\sigma_n^{2k} \quad \text{where} \quad \sigma_n^2 = \text{var } x_{jn}.$$

If we normalize $w(t)$ so $\text{var } w(t) = t$, then $\sigma_n^2 = \frac{t}{n}$ and $\text{var } I_k(n) = \binom{n}{k}\frac{t^k}{n^k} \to \frac{t^k}{k!}$ as $n \to \infty$. We put $I_k(n) \to I_k(t)$ as $n \to \infty$, which is the kth iterated stochastic integral of the process $w(t)$.

We have $$E_n = \prod_1^n (1+vx_{jn}).$$

Proposition 46: (1) E_n converges in distribution to E, satisfying

$$\log\langle E^z\rangle = L(D)(1+vx)^z\big|_{x=0}$$

(2) The distributions and joint distributions of the $I_k(n)$ converge to those of $I_k(t)$ such that $E(v) = \Sigma\, v^k I_k(t)$.

(3) The joint distributions of $E(v_1),\dots,E(v_N)$ satisfy

$$\langle \prod_1^N E(v_j)^{z_j}\rangle = \exp\zeta_N \quad \text{where} \quad \zeta_N = L(D)\prod_1^N (1+v_jx)^{z_j}\big|_{x=0}.$$

Proof: (1) We have $\langle E_n{}^z\rangle = \prod_1^n \langle (1+vx_{jn})^z\rangle = \prod_1^n (\Sigma \frac{z^{(k)}}{k!} v^k \langle x_{jn}{}^k\rangle)$.

Now note that, for $X = x_{jn}$,

$$\langle X^k\rangle = \langle w(\tfrac{t}{n})^k\rangle = \partial_z^k\big|_0 e^{\frac{t}{n}L(z)} .$$

We have $e^{\frac{t}{n}L(z)} = 1 + \frac{t}{n}L(z) + \frac{t^2}{2n^2}g(z)$ where $g(z)$ is analytic in a neighborhood of $0 \in \mathbb{C}$ and $g(0) = 0$. We thus have, for $k \geq 1$,

$$\langle X^k\rangle = \frac{t}{n}L^{(k)}(0) + \frac{t^2}{2n^2}\, g^{(k)}(0).$$

Therefore,

$$\Sigma \frac{z^{(k)}}{k!} v^k \langle X^k\rangle = 1 + \frac{t}{n}\, \Sigma \frac{z^{(k)}}{k!} v^k L^{(k)}(0) + \frac{t^2}{2n^2}\, \Sigma \frac{z^{(k)}}{k!} v^k g^{(k)}(0).$$

And the limit of $\langle E_n{}^z\rangle = \lim_{n\to\infty} t(1+\frac{t}{n}\zeta)^n = e^{t\zeta}$ where

$$\zeta = \Sigma \frac{z^{(k)}}{k!}v^k L^{(k)}(0) = \Sigma \frac{z^{(k)}}{k!}v^k\ \Sigma\binom{k}{\ell}x^{k-\ell}L^{(\ell)}(0)\big|_{x=0}$$

$$= \Sigma \frac{z^{(k)}v^k}{k!}L(D)x^k\big|_{x=0} \quad \text{(by GLM)}$$

$$= L(D)(1+vx)^z\big|_{x=0}.$$

(2) and (3): To check convergence of the I's we have to check the joint convergence of $\langle E_n(v_1)^{z_1}\ldots E_n(v_n)^{z_N}\rangle$ for every $1 \leq N < \infty$.

We have $\langle \prod_1^N E_n(v_j)^{z_j}\rangle = \langle \prod_{j=1}^N (1+v_jX_n)^{z_j}\rangle^n$ where X_n has the same distribution as x_{jn}. And, using the notation $|k| = \Sigma k_j$ for multi-indices,

$$\langle \prod_{j=1}^N (1+v_jX_n)^{z_j}\rangle = \Sigma\ldots\Sigma \prod_{j=1}^N \frac{z_j^{(k_j)}}{k!}v_j^{k_j}\langle X_n{}^{|k|}\rangle$$

$$= 1 + \sum_{k\neq 0}\ldots\Sigma \prod_{j=1}^N \frac{z_j^{(k_j)}}{k_j!}v_j^{k_j}\frac{t}{n}L^{(|k|)}(0) + \frac{t^2}{2n^2}\text{(rem.)}$$

$$= 1 + \frac{t}{n}\zeta_N + \Theta(n^{-2}).$$

So the joint limit is $e^{t\zeta_N}$, where, as for (1),

$$\zeta_N = L(D)\prod_{j=1}^N (1+v_jx)^{z_j}\big|_{x=0}.$$

Remarks:

(1) To check orthogonality of the $I_k(t)$ we have to see that $\langle E(v_1)(E(v_2)\rangle$ is a function only of v_1v_2. This is the case N=2, z=1, and we have, for t=1,

$$\langle E(v_1)E(v_2)\rangle = \exp[L(D)(1+v_1x)(1+v_2x)\big|_{x=0}]$$

$$= \exp[L(0) + (v_1+v_2)L'(0) + v_1v_2L''(0)]$$

$$= e^{v_1v_2} \quad \text{since we have normalized } L(0) = L'(0) = 0,\ L''(0) = 1.$$

Observe, then, that this procedure yields an orthogonal sequence for any L(D). We also have $\langle I_k{}^2(t)\rangle = \frac{t^k}{k!}$ as seen above.

(2) By the duality lemma,

$$L(D)(1+vx)^z|_{x=0} = (1+v\partial_a)^z e^{ax}L(a)|_{a=0,x=0} = (1+v\partial_D)^z L(0).$$

Examples. (1) $L(D) = \frac{D^2}{2}$. We have $\zeta = \frac{1}{2}z^{(2)}v^2$ and

$$\langle E^z\rangle = e^{\frac{t}{2}(z^2v^2-zv^2)} = \langle (e^{vb(t)-\frac{v^2t}{2}})^z\rangle$$ for Brownian motion

$b(t)$. We check the joint distribution of $E(v_j)$, $1 \leq j \leq N$.

$$\langle \Pi E(v_j)^{z_j}\rangle = \exp[\tfrac{t}{2}(\Sigma z_jv_j)^2 - \tfrac{t}{2}\Sigma z_jv_j^2].$$

From Proposition 46 we have

$$\zeta_N = \tfrac{1}{2}D^2 \prod_1^N (1+v_jx)^{z_j}|_{x=0}$$

$$= \tfrac{1}{2}(\Sigma \frac{z_jv_j}{1+v_jx})^2 - \tfrac{1}{2}\Sigma \frac{z_jv_j^2}{(1+v_jx)^2}|_{x=0}$$

$$= \tfrac{1}{2}(\Sigma z_jv_j)^2 - \tfrac{1}{2}\Sigma z_jv_j^2 \quad \text{as expected.}$$

$E(v) = G[b(t),t;v]$, where G is the generating function of the Hermite polynomials, the appropriate orthogonal system as we saw in Chapter IV.

2. $L(D) = e^D-1-D$. We have $\zeta = (e^D-1-D)(1+vx)^z|_{x=0} = (1+v)^z-1-vz$.

So $\langle E^z\rangle = \langle [e^{-vt}(1+v)^{N(t)}]^z\rangle$ for the Poisson process $N(t)$. Again we check the joint distributions.

$$\langle \Pi\, E(v_j)^{z_j}\rangle = \exp[t(-\Sigma v_jz_j + \prod_1^N (1+v_j)^{z_j}-1)].$$

From Proposition 46,

$$\zeta_N = (e^D-1-D)\Pi(1+v_jx)^{z_j}|_{x=0} = \Pi(1+v_j)^{z_j}-1-\Sigma v_jz_j \quad \text{in agreement.}$$

We see that E is $G[N(t),t;v]$ as found in Chapter IV so that $I_k(t) = \frac{1}{k!}P_k$, P_k the Poisson-Charlier polynomials.

The problem in general, however, to determine $E(v)$ from the conditions

$$\langle \Pi E(v_j)^{z_j}\rangle = e^{t\zeta_N},$$

is not easy. We state the results of these examples as the following:

Theorem 4 1. Let x_{jn} be independent, Gaussian, with mean zero, variance t/n.

Let $I_k(n) = k^{th}$ elementary symmetric polynomial in the $\{x_{jn}\}$. Then $k!I_k(n)$ converge in distribution to $H_k[b(t),t]$, the kth Hermite polynomial, $b(t)$ Brownian motion.

2. For $x_{jn}+\frac{t}{n}$ independent Poisson with mean t/n, $k!I_k(n)$ converge in distribution to $P_k[N(t),t]$, the kth Poisson-Charlier polynomial, $N(t)$ the standard Poisson process.

We now discuss approximation by simpler processes. First we have a general limit result.

Proposition 47: Let $w_n(t)$ be generated by $L_n(z)$, analytic in some fixed neighborhood of $0 \in \mathbb{C}$. Then if $L_n(z) \to L(z)$ uniformly on compact subsets of $\mathbb{C}$, $E_n(v)$ and $I_k^n(t)$ corresponding to $w_n(t)$ converge in distribution to $E(v)$, $I_k(t)$ corresponding to $w(t)$.

Proof: The joint distribution of $\{E_n(v_j)\}_{1\le j\le N}$ is determined by

$$\zeta_N^n = L_n(D)\prod_1^N(1+v_jx)^{z_j}\Big|_{x=0}$$

which depends only on derivatives of L_n at 0. Since $L_n \to L$ normally, the derivatives at 0 converge, e.g. by Cauchy's integral formulas, which also show that they grow at most geometrically ($L_n^{(k)} \sim \mathcal{C}^k$) uniformly in n. Thus

$$\zeta_N^n \to \zeta_N = L(D)\prod_1^N(1+v_jx)^{z_j}\Big|_{x=0}.$$

We consider a classical case of the Central Limit Theorem.

Proposition 48: Let $x_{jn} = \frac{x_j}{\sqrt{n}}$ where x_j are independent Bernoulli variables, $P(x_j = \pm 1) = \frac{1}{2}$. Let $E_n = \prod_1^n(1+vx_{jn})$, $I_k(n) = k^{th}$ elementary symmetric polynomial in $\{x_{jn}\}$. Then as $n \to \infty$,

$$E_n \to e^{vb(1)-v^2/2}, \quad k!I_k(n) = H_k[b(1),1].$$

Proof: We check $\langle\prod_1^N E_n(v_\ell)^{z_\ell}\rangle = [\frac{1}{2}\prod_1^N(1+\frac{v_\ell}{\sqrt{n}})^{z_\ell} + \frac{1}{2}\prod_1^N(1-\frac{v_\ell}{\sqrt{n}})^{z_\ell}]^n$

$$= \Big(\tfrac{1}{2}\prod_1^N[1+\frac{z_\ell v_\ell}{\sqrt{n}}+\frac{z_\ell^{(2)}v_\ell^2}{2n}+o(n^{-1})] + \tfrac{1}{2}\prod_1^N[\ -\ ]\Big)^n$$

$$\approx \left(1 + \frac{2}{2n} \sum_{\ell<m}\sum z_\ell v_\ell z_m v_m + \frac{1}{2n} \Sigma z_\ell^{(2)} v_\ell^2\right)^n$$

$$= \left[1 + \frac{1}{2n}(\Sigma z_\ell v_\ell)^2 - \frac{1}{2n} \Sigma z_\ell v_\ell^2\right]^n$$

$$\to \exp\left[\tfrac{1}{2}(\Sigma z_\ell v_\ell)^2 - \tfrac{1}{2} \Sigma z_\ell v_\ell^2\right].$$

In view of Theorem 4, we can say that $x_{jn} \to db$ as $n \to \infty$. For the Poisson process we have, similarly,

Proposition 49: Let x_{jn} be independent Bernoulli variables, $P(x_{jn} = 0) = 1 - \frac{1}{n}$, $P(x_{jn} = 1) = \frac{1}{n}$.

Let $E_n = \prod_1^n (1+vx_{jn})$, $I_k(n) = k^{th}$ symmetric polynomial in $\{x_{jn}\}$.

Then, as $n \to \infty$,

$$E_n \to (1+v)^{N(1)}, \quad k!I_k(n) \to N^{(k)}, \quad N = N(1).$$

Proof: We have

$$\langle \prod_1^m E_n(v_\ell)^{z_\ell} \rangle = \left[\frac{1}{n} \prod_1^m (1+v_\ell)^{z_\ell} + 1 - \frac{1}{n}\right]^n$$

$$\to \exp\left[\prod_1^m (1+v_\ell)^{z_\ell} - 1\right] = \langle \prod_1^m (1+v_\ell)^{z_\ell N} \rangle.$$

So $$E = (1+v)^N = \Sigma \frac{N^{(k)}}{k!} v^k.$$

Note that here we do not get an orthogonal system; $L = e^D - 1$ here, so $L'(0) \neq 0$.

We do get the iterated integrals of dN since as we saw in Chapter V,

$$N^{(k)} \text{ satisfies } \int_0^t N^{(k)} dN = \frac{1}{k+1} N^{(k+1)}, \quad \partial_1^+ N^{(k)} = kN^{(k-1)}.$$

We can thus say in this case that $x_{jn} \to dN$ as $n \to \infty$. We state the results of the probabilistic theory.

Theorem 5. Let $w(t)$ be a process with independent increments generated by $L(D)$. Then the functional $E(v)$ satisfying

$$\langle E(v)^z \rangle = \exp[t\, L(D)(1+vx)^z|_{x=0}]$$

and similarly for the joint distributions of $E(v_1), \ldots, E(v_N)$, is the generating function for the iterated stochastic integrals of $dw(t)$.

PART 3. EXTENSIONS

CHAPTER VIII. MULTIDIMENSIONAL THEORY

We will study the moment theory and orthogonal theory for vector processes $\underset{\sim}{w}(t)$.

MOMENT THEORY

For a process in R^N, $\underset{\sim}{w}(t)$, we have the relation $\langle e^{\underset{\sim}{z}\cdot\underset{\sim}{w}(t)}\rangle = e^{tL(\underset{\sim}{z})}$.

The variables x_j and $z_j = D_j = \frac{\partial}{\partial x_j}$ satisfy the commutation relations $[x_j,x_k] = [z_j,z_k] = 0$, $[z_j,x_k] = \delta_{jk}$.

The operator $\underset{\sim}{C} = \underset{\sim}{x}(t) = \underset{\sim}{x} + t\underset{\sim}{\nabla}L$ since we have $x_j(t) = x_j + t\frac{\partial L}{\partial D_j}$.

The operator $\underset{\sim}{A} = (C_1D_1,\dots,C_ND_N)$ has an associated scalar operator $A = \underset{\sim}{C}\cdot\underset{\sim}{D} = \Sigma\, C_jD_j$ which is our <u>number</u> <u>operator</u>.

Define $h_{\underset{\sim}{n}} = C_1^{n_1}\dots C_N^{n_N}1$.

We use the multi-index notations $|\underset{\sim}{n}| = \Sigma n_j$, $\underset{\sim}{x}^{\underset{\sim}{n}} = \prod_1^N x_j^{n_j}$, $\underset{\sim}{n}! = \prod n_j!$.

<u>Proposition 50</u>: (1) $e^{\underset{\sim}{z}\cdot\underset{\sim}{C}}1 = \Sigma\, \frac{\underset{\sim}{z}^{\underset{\sim}{n}}}{\underset{\sim}{n}!}\, h_{\underset{\sim}{n}}(\underset{\sim}{x},t)$

(2) $C_jh_{\underset{\sim}{n}} = h_{\underset{\sim}{n}+\underset{\sim}{e}_j}$, where $\underset{\sim}{e}_j = (0,\dots,1,\dots,0)$ in j^{th} place.

(3) $D_jh_{\underset{\sim}{n}} = n_jh_{\underset{\sim}{n}-\underset{\sim}{e}_j}$

(4) $Ah_{\underset{\sim}{n}} = |\underset{\sim}{n}|h_{\underset{\sim}{n}}$

<u>Proof</u>: (1)-(4) follow from properties for $N = 1$.

Proposition 5 becomes, for $H = H(\underset{\sim}{x},\underset{\sim}{z})$,

(1) $[x_j(t),x_k(t)] = [z_j(t),z_k(t)] = 0$, $[z_j(t),x_k(t)] = \delta_{jk}$.

(2) $\dot{x}_j(t) = \frac{\partial H}{\partial z_j}[\underset{\sim}{x}(t),\underset{\sim}{z}(t)]$.

(3) $\dot{z}_j(t) = -\frac{\partial H}{\partial x_j}[\underset{\sim}{x}(t),\underset{\sim}{z}(t)]$.

Note that the C_j's and A_j's commute among themselves. Vacuum functions are much more interesting in the case $N > 1$. We first note

<u>Proposition 51</u>: For a function $\Omega(\underset{\sim}{x},t)$, set $\Omega_{\underset{\sim}{n}} = \underset{\sim}{C}^{\underset{\sim}{n}}\Omega$.

(1) $\underset{\sim}{A}\Omega = 0$ implies $A\Omega_{\underset{\sim}{n}} = |\underset{\sim}{n}|\Omega_{\underset{\sim}{n}}$.

(2) $A\Omega = 0$ implies $A\Omega_{\underset{\sim}{n}} = |\underset{\sim}{n}|\Omega_{\underset{\sim}{n}}$.

<u>Proof</u>: (1) follows as for $N = 1$.

(2) $A\Omega_{\underset{\sim}{n}} = \Sigma\, x_j(t)z_j(t) \prod_1^N x_\ell^{n_\ell}(t)\Omega$

$= |\underset{\sim}{n}|\Omega_{\underset{\sim}{n}} + \Sigma \Pi\, x_\ell^{n_\ell}(t)x_j(t)z_j(t)\Omega$

$= |\underset{\sim}{n}|\Omega_{\underset{\sim}{n}} + \underset{\sim}{C}^{\underset{\sim}{n}}A\Omega.$

We thus make the

<u>Definition</u>: Ω is a <u>vacuum function</u> if it satisfies

(1) Ω is coharmonic.

(2) $A\Omega = \Sigma A_j\Omega = 0$.

(3) Ω is <u>absolute</u> if $\frac{\partial\Omega}{\partial t} = 0$.

Theorem 1 and Proposition 51 imply, then, that if Ω is a vacuum function, $\Omega_{\underset{\sim}{n}}$ is coharmonic and $A\Omega_{\underset{\sim}{n}} = |\underset{\sim}{n}|\Omega_{\underset{\sim}{n}}$.

<u>Proposition 52</u>: At $t = 0$, $A = A(0) = \underset{\sim}{x}\cdot\underset{\sim}{D}$. Thus, $A(0)\Omega = 0$ if and only if Ω is homogeneous of degree zero.

<u>Proof</u>: (1) $\lambda^{\underset{\sim}{x}\cdot\underset{\sim}{D}}\Omega = \Omega(\lambda\underset{\sim}{x})$. Apply $\frac{\partial}{\partial\lambda}$ to get $0 = \frac{\partial}{\partial\lambda}\Omega(\lambda\underset{\sim}{x})$, i.e. $\Omega(\lambda\underset{\sim}{x})$ is independent of λ.

(2) $\Omega(\lambda\underset{\sim}{x}) = \Omega(\underset{\sim}{x})$ yields, as above, $\underset{\sim}{x}\cdot\underset{\sim}{D}\Omega = 0$.

<u>Remarks</u>: So we have that an absolute vacuum is any harmonic function $\Omega(\underset{\sim}{x})$, homogeneous of degree zero. Our canonical vacuums are still 1 and $e^{tL}\chi(x_1)\ldots\chi(x_N)$; but we now have many functions and, hence, corresponding series. A "general solution" to $\underset{\sim}{x}\cdot\underset{\sim}{D}h = 0$ is

$$h = \sum_{\substack{\underset{\sim}{a} \in R^N \\ \underset{\sim}{a} \cdot \underset{\sim}{1} = 0}} Q_{\underset{\sim}{a}}(\underset{\sim}{x}^{\underset{\sim}{a}}) \quad \text{where } Q_{\underset{\sim}{a}} \text{ are functions of one variable.}$$

INDEPENDENT PROCESSES

If $\underset{\sim}{w}(t) = [w_1(t),\ldots,w_N(t)]$, where the $w_j(t)$ are independent processes, then the multidimensional generator is simply the sum of the generators of the w_j's. That is, for

$$\underset{\sim}{W} = [w_1(t_1),\ldots,w_N(t_N)], \quad \langle e^{\underset{\sim}{z}\cdot\underset{\sim}{W}}\rangle = \prod_1^N \langle e^{z_j w_j(t_j)}\rangle = e^{\Sigma t_j L_j(z_j)}.$$

Setting $t_j = t$, $1 \le j \le N$, we have $L(\underset{\sim}{z}) = \Sigma L_j(z_j)$.

<u>Proposition 53</u>: If the components $w_j(t)$ are independent, with corresponding generators $L_j(z_j)$, then

(1) $C_j = x_j + tL_j'(D_j)$

(2) $e^{\underset{\sim}{z}\cdot\underset{\sim}{C}}1 = \prod_1^N (e^{z_j C_j}1)$

(3) $h_{\underset{\sim}{n}}(\underset{\sim}{x},t) = \prod_1^N h_{n_j}(x_j,t)$

<u>Proof</u>: (1) $C_j = x_j + t\frac{\partial L}{\partial D_j} = x_j + tL_j'(D_j)$

(2) $e^{\underset{\sim}{z}\cdot\underset{\sim}{C}}1 = e^{tL}e^{\underset{\sim}{z}\cdot\underset{\sim}{x}} = \Pi\, e^{tL_j}e^{\Sigma z_\ell x_\ell} = e^{\underset{\sim}{z}\cdot\underset{\sim}{x}}e^{t\Sigma L_j(z_j)} = \prod_1^N (e^{z_j C_j}1).$

(3) Expand the result of (2) to get $\Sigma \frac{\underset{\sim}{z}^{\underset{\sim}{n}}}{\underset{\sim}{n}!} h_{\underset{\sim}{n}} = \prod_j \sum_{n_j} \frac{z_j^{n_j}}{n_j!} h_{n_j}.$

<u>Examples</u>: (1) Brownian Motion. $L = \frac{1}{2}\Sigma z_j^2$. Harmonic functions are the usual harmonic functions. Since any harmonic f, in the classes of Proposition 1, is analytic, the only absolute vacuums are constant functions. The polynomials

$$\overline{h}_{\underset{\sim}{n}}(\underset{\sim}{x},t) = \prod_1^N H_{n_j}(x_j,t),$$

products of Hermite polynomials.

(2) Consider the wave operator $L = D_{x_1}D_{x_2}$. We put $x_1 = x$, $x_2 = y$.

We have $\underset{\sim}{C} = (x+tD_y, y+tD_x)$. The moment polynomials are thus

$$h_{nm} = (x+tD_y)^n(y+tD_x)^m 1$$
$$= (x+tD_y)^n y^m$$
$$= \Sigma \frac{m^{(k)}n^{(k)}}{k!} y^{m-k}t^k x^{n-k}.$$

The number operator $A = xD_x+yD_y+2tD_xD_y$. $Ah_{nm} = (n+m)h_{nm}$.

A harmonic function, independent of t, is of the general form $f = F(x)+G(y)$. To be a vacuum, $(xD_x+yD_y)f = 0 = xF'(x)+yG'(y)$, that is, $xF'(x) = \text{constant} = -yG'(y)$. We choose the constant to be 1, $F(x) = \log x$, $G(y) = -\log y$ and $f = \log\frac{x}{y}$, homogeneous of degree zero. We thus have

$$u = e^{zC_x+wC_y}f = e^{zx+ztD_y}e^{wy+wtD_x}\log\frac{x}{y} = e^{zx}e^{wy+wzt}\log\frac{x+wt}{y+zt}.$$

We have $f_{nm} = C_x^n C_y^m(\log x-\log y)$. We first have, using GLM,

$$C_x^n C_y^m \log x = (x+tD_y)^n(y+tD_x)^m \log x$$
$$= (x+tD_y)^n[(\log x)y^m + \sum_{k\geq 1} \frac{(-1)^{k+1}\Gamma(k)}{x^k}\binom{m}{k}y^{m-k}t^k]$$
$$= h_{nm}\log x + \sum_{k\geq 1} \frac{(-1)^{k+1}m^{(k)}}{kx^k} t^k h_{n,m-k}.$$

Similarly,

$$C_x^n C_y^m \log y = h_{nm}\log y + \sum_{k\geq 1} \frac{(-1)^{k+1}n^{(k)}}{ky^k} t^k h_{n-k,m}.$$

Thus, $f_{nm} = h_{nm}\log\frac{x}{y} + \sum_{k\geq 1} \frac{(-1)^{k+1}}{k} t^k[\frac{m^{(k)}}{x^k} h_{n,m-k} - \frac{n^{(k)}}{y^k} h_{n-k,m}]$.

u and f_{nm} satisfy $\frac{\partial F}{\partial t} = D_xD_yF$; $Af_{nm} = (n+m)f_{nm}$.

(3) We can consider a general wave operator $H = D_y^2-L^2$ where L is a function of $\underset{\sim}{D} = (D_1,\dots,D_N)$, $D_j = \frac{\partial}{\partial x_j}$. We have

$$C_j = x_j - 2tL\frac{\partial L}{\partial D_j} = x_j - 2tLL_j. \quad C_y = y + 2tD_y.$$

The number operator is

$$A = (y,\underset{\sim}{x})\cdot(D_y,\underset{\sim}{D}) + 2t(D_y^2-L\underset{\sim}{\nabla}L\cdot\underset{\sim}{D}), \quad \text{where } \underset{\sim}{\nabla}L = (\frac{\partial L}{\partial D_1},\dots,\frac{\partial L}{\partial D_N}).$$

We choose an absolute vacuum $\Omega_\pm$ to satisfy $\frac{\partial\Omega_\pm}{\partial y} \pm L\Omega_\pm = 0$, i.e.

$\Omega_\pm = e^{\pm yL}\Omega_o(\underset{\sim}{x})$. We need $\Omega_\pm$ to be homogeneous of degree zero in $(y,\underset{\sim}{x})$. That is,

$$\Omega_\pm(\lambda y,\lambda\underset{\sim}{x}) = \Omega_\pm(y,\underset{\sim}{x}) = \lambda^{\underset{\sim}{x}\cdot\underset{\sim}{D}} e^{\pm\lambda yL}\Omega_o(\underset{\sim}{x}).$$

If we put $y = 0$, we have Ω_o homogeneous of degree zero. We now use the formula $g(D)\lambda^{xD}f(x) = \lambda^{xD}g(\lambda D)f(x)$ (see Proposition 15), which extends to higher dimensions. This yields, putting $\underset{\sim}{D} \to \lambda^{-1}\underset{\sim}{D}$ in the exponent,

$$\Omega_\pm(y,\underset{\sim}{x}) = e^{\pm\lambda yL(\lambda^{-1}D)}\lambda^{\underset{\sim}{x}\cdot\underset{\sim}{D}}\Omega_o(\underset{\sim}{x}).$$

Since $\lambda^{\underset{\sim}{x}\cdot\underset{\sim}{D}}\Omega_o = \Omega_o$ we must have $\lambda L(\lambda^{-1}D) = L(D)$, i.e. L must be homogeneous of degree 1 in order for H to admit non-constant vacuum functions. E.g. we could choose

$L = -\sqrt{-\underset{\sim}{D}^2} = -\sqrt{-\sum_1^N D_j^2}$. This generates the Cauchy process on R^N and we have, for $y > 0$, $\underset{\sim}{v}^2 = \underset{\sim}{v}\cdot\underset{\sim}{v}$,

$$e^{yL}f(\underset{\sim}{x}) = \frac{\Gamma[(N+1)/2]}{\pi^{(N+1)/2}}\int_{R^N}\frac{1}{(1+\underset{\sim}{v}^2)^{(N+1)/2}}f(\underset{\sim}{x}+y\underset{\sim}{v})d\underset{\sim}{v}.$$

$H = D_y^2+\underset{\sim}{D}^2$ = Laplacian in $(y,\underset{\sim}{x})$. $e^{yL}f(\underset{\sim}{x}) = \Omega$ satisfies $H\Omega = 0$.

<u>Proposition 54</u>: Let $H = D_y^2-L^2$ where $L(D_1,\dots,D_N)$ is homogeneous of degree one. Let $\Omega_o(\underset{\sim}{x})$ be homogeneous of degree zero. Then

(1) $\Omega = e^{yL}\Omega_o(\underset{\sim}{x})$ is a vacuum function.

(2) $C_y\Omega = (y+2tL)\Omega$.

Let $\Omega_{n\underset{\sim}{m}} = C_y^n\underset{\sim}{C}^{\underset{\sim}{m}}\Omega$.

(3) $$\begin{aligned}A\Omega_{n\underset{\sim}{m}} &= [(y,\underset{\sim}{x})\cdot(D_y,\underset{\sim}{D}) + 2tH]\Omega_{n\underset{\sim}{m}}\\ &= (n+|\underset{\sim}{m}|)\Omega_{n\underset{\sim}{m}}.\end{aligned}$$

<u>Proof</u>: (1) follows from above.

(2) $\Omega = e^{yL}\Omega_o$ so that $D_y\Omega = L\Omega$. $C_y = y+2tD_y$.

(3) Since L is homogeneous of degree 1, applying $\left.\frac{\partial}{\partial\lambda}\right|_{\lambda=1}$ to $\lambda^{\underset{\sim}{D}\cdot\underset{\sim}{\nabla}}L = \lambda L$, $\underset{\sim}{D}\cdot\underset{\sim}{\nabla}L = L$ (Euler's Theorem).

As noted above,

$$A = yDy + \underset{\sim}{x}\cdot\underset{\sim}{D} + 2t(D_y^2 - L\underset{\sim}{D}\cdot\underset{\sim}{\nabla}L).$$

And

$$D_y^2 - L\underset{\sim}{D}\cdot\underset{\sim}{\nabla}L = D_y^2 - L^2 = H.$$

RADIAL AND ANGULAR VARIABLES

We can express functions on R^N in terms of $r = |\underset{\sim}{x}|$ and $\underset{\sim}{u}$ = points on the unit sphere S^{N-1}. Denote by θ_j the angle($\underset{\sim}{x}, x_j$ axis), $u_j = \cos\theta_j$; and introduce the angular momentum operators $a_{jk} = x_jD_k - x_kD_j = \frac{d}{d\theta_{jk}}$, where θ_{jk} is measured from the x_j-axis in the (x_j, x_k) plane.

It is readily checked that $a_{jk} = -a_{kj}$ and that for $j<k$, $\ell<m$, $[a_{jk}, a_{\ell m}] = a_{jm}\delta_{k\ell}$.

We see that $\underset{\sim}{x} = r\underset{\sim}{u}$, $\frac{\partial}{\partial r} = \underset{\sim}{u}\cdot\underset{\sim}{D}$ and $[\frac{\partial}{\partial r}, a_{jk}] = 0$.

We also note that the radius vector from the origin to the point $\underset{\sim}{x}$ is normal to S^{N-1} so that r and $\underset{\sim}{u}$ are independent coordinates.

In the $\underset{\sim}{z}$-space, momentum space, we denote $|\underset{\sim}{z}|$ by ρ, $\underset{\sim}{z} = \rho\underset{\sim}{\omega}$, and $z_j\partial_{z_k} - z_k\partial_{z_j}$ by $\alpha_{jk} = \frac{\partial}{\partial\psi_{jk}}$.

<u>Proposition 55</u>: (1) $\underset{\sim}{x}\cdot\underset{\sim}{D} = r\frac{\partial}{\partial r}$

(2) If f is homogeneous of degree zero, f is a function of $\underset{\sim}{u}$, $f(\underset{\sim}{u})$.

(3) If f is homogeneous of degree p, f is of the form $r^pg(\underset{\sim}{u})$.

<u>Proof</u>: (1) $\frac{\partial}{\partial r} = \frac{\partial\underset{\sim}{x}}{\partial r}\cdot\underset{\sim}{D} = \underset{\sim}{u}\cdot\underset{\sim}{D} = \frac{1}{r}\underset{\sim}{x}\cdot\underset{\sim}{D}$.

(2) $f(\lambda\underset{\sim}{x}) = \lambda^{\underset{\sim}{x}\cdot\underset{\sim}{D}}f(\underset{\sim}{x}) = \lambda^{r(\partial/\partial r)}f(r,\underset{\sim}{u}) = f(\lambda r,\underset{\sim}{u}) = f(\underset{\sim}{x})$. Let $\lambda\to 0$.

(3) $f(\lambda\underset{\sim}{x}) = f(\lambda r,\underset{\sim}{u}) = \lambda^pf(\underset{\sim}{x})$. Set $\lambda = \frac{1}{r}$. Then we have,

$f(\underset{\sim}{x}) = r^pf(1,\underset{\sim}{u})$.

<u>Corollary</u>: Ω is an absolute vacuum if and only if $H\Omega = 0$ and Ω is a function only of $\underset{\sim}{u}$.

(3) implies that the generators L homogeneous of degree one must be of the form $\sqrt{\underset{\sim}{D}^2}\, g(\underset{\sim}{\omega})$.

We note that in general the number operator is $e^{tH}r\frac{\partial}{\partial r}e^{-tH}$.

We now see that

Proposition 56: For $H = L(z)$, $A = r\frac{\partial}{\partial r} + t\rho\frac{\partial L}{\partial \rho}$.

Proof: $A = \underset{\sim}{x}\cdot\underset{\sim}{D} + t(\underset{\sim}{D}\cdot\underset{\sim}{\nabla})L$. From Proposition 55, $\underset{\sim}{x}\cdot\underset{\sim}{D} = r\frac{\partial}{\partial r}$, $\underset{\sim}{z}\cdot\underset{\sim}{\nabla}_z = \rho\frac{\partial}{\partial \rho}$.

SOME SIMPLE INVARIANCE PROPERTIES

In general if the generators of a Lie algebra commute with H, then they will be invariant under the conjugation by e^{-tH}. The group they generate is thus a symmetry group of the flow. E.g. translation invariance is a property of $H = L(\underset{\sim}{z})$, since $[z_j, H] = 0$, $1 \le j \le N$. We discuss some basic examples.

1. Number operator. We call this scale invariance. We ask when $\underset{\sim}{x}(t)\cdot\underset{\sim}{z}(t) = \underset{\sim}{x}\cdot\underset{\sim}{z}$. By Propositions 56 and 55, for $H = L(\underset{\sim}{z})$, we must have $\rho\frac{\partial L}{\partial \rho} = 0$, or L is homogeneous of degree zero, $L = L(\underset{\sim}{\omega})$. For general H we have

Proposition 57: H is scale invariant if $\Sigma\frac{\partial H}{\partial z_j}z_j = \Sigma x_j\frac{\partial H}{\partial x_j}$.

Proof: $[H, \Sigma x_j z_j] = \Sigma Hx_j z_j - \Sigma x_j z_j H = \Sigma\frac{\partial H}{\partial z_j}z_j - \Sigma x_j\frac{\partial H}{\partial x_j}$.

Under such a flow, then, homogeneity properties of functions are preserved; we have $\lambda^{xD}f[x(t)] = e^{tH}\lambda^{xD}e^{-tH}e^{tH}f(x)e^{-tH} = f[\lambda x(t)]$.

2. Divergence. $\text{div} = \underset{\sim}{1}\cdot\underset{\sim}{z}$. We see that $\text{div}(t) = \text{div}$ if $\Sigma\frac{\partial H}{\partial x_j} = 0$, i.e. $\text{div}_{\underset{\sim}{x}}H = 0$.

3. Curl. For R^3, $\text{curl} = \underset{\sim}{z}\times$. $\text{curl}(t) = \text{curl}$, then, just in the case H is translation-invariant.

4. Angular momentum. For R^3, e.g., $\underset{\sim}{a} = \underset{\sim}{x}\times\underset{\sim}{z} = (\frac{\partial}{\partial\theta_{23}}, \frac{\partial}{\partial\theta_{31}}, \frac{\partial}{\partial\theta_{12}})$. If $\underset{\sim}{a}(t) = \underset{\sim}{a}$, then H is rotationally invariant.

   We set $\gamma_{jk}F = \frac{\partial F}{\partial z_j}z_k - \frac{\partial F}{\partial z_k}z_j$, for an operator $F(\underset{\sim}{x}, \underset{\sim}{z})$.

   For F depending only on $\underset{\sim}{z}$, $\gamma_{jk} = -\alpha_{jk}$. We have

Proposition 58: H is rotationally invariant if $a_{jk}H = \gamma_{jk}H$ for $1 \le j < k \le 3$. In particular, if $H = L(\underset{\sim}{z})$, H is rotationally invariant if and only if $L = L(\rho)$.

<u>Proof</u>: Consider $a_{12} = x\partial_y - y\partial_x \equiv xw - yz$.

Then

$$[H,a_{12}] = Hxw - Hyz + yzH - xwH$$
$$= \frac{\partial H}{\partial z}w - \frac{\partial H}{\partial w}z + y\frac{\partial H}{\partial x} - x\frac{\partial H}{\partial y}$$
$$= (\gamma_{jk} - a_{jk})H.$$

In case $H = L(\underset{\sim}{z})$, $a_{jk}L = 0$ and $\therefore\ \alpha_{jk}L = 0$.

We have $\alpha_{jk} = \frac{d}{d\psi_{jk}}$ where ψ_{jk} is the angle with the z_j-axis of the projection of $\underset{\sim}{z}$ onto the (z_j,z_k) plane. From $\omega_j = \frac{z_j}{\rho}$ we have,

$$\psi_{jk} = \arctan\frac{z_k}{z_j} = \arctan\frac{\omega_k}{\omega_j}, \quad \text{i.e. } \omega_k = \omega_j\tan\psi_{jk}.$$

Since $|\underset{\sim}{\omega}| = 1$, $\omega_j = \pm\frac{1}{\sqrt{1 + \sum_k \tan^2\psi_{jk}}}$.

Thus, by adjusting the ψ_{jk} we can vary the ω_j.

From $\frac{\partial L}{\partial\omega_j} = \sum_{k\neq j}\frac{\partial L}{\partial\psi_{jk}}\frac{\partial\psi_{jk}}{\partial\omega_j} = 0$ we conclude $L = L(\rho)$.

5. Energy. This is the most immediate. For $H = H(\underset{\sim}{x},\underset{\sim}{z})$,

$$H(t) = H[\underset{\sim}{x}(t),\underset{\sim}{z}(t)] = H(\underset{\sim}{x},\underset{\sim}{z}),$$

i.e. H itself is invariant, since, of course, $[H,H] = 0$.

6. Time-dependent Operators. As we noted in Chapter II in the section on vacuum functions, a harmonic operator $U(\underset{\sim}{x},\underset{\sim}{z},t)$ evaluated along the flow is constant, i.e. $U[\underset{\sim}{x}(t),\underset{\sim}{z}(t),t] = U(\underset{\sim}{x},\underset{\sim}{z},0)$. Time-independent harmonic operators are just those commuting with H, as above.

<u>Examples</u>: Let's consider $H = A = x\cdot D = r\frac{\partial}{\partial r}$.

(1) We check scale invariance:

$$\sum\frac{\partial H}{\partial z_j}z_j = \sum x_jz_j = \sum x_j\frac{\partial H}{\partial x_j}, \qquad \underset{\sim}{x}(t)\cdot\underset{\sim}{z}(t) = \underset{\sim}{x}\cdot\underset{\sim}{z} = H,$$

so this is also invariance of H.

(2) We can calculate explicitly:

$$\dot{x}_k(t) = x_k(t) \quad \text{so} \quad x_k(t) = x_ke^{t}$$
$$\dot{z}_k(t) = -z_k(t) \quad \text{so} \quad z_k(t) = z_ke^{-t}.$$

(3) $\mathrm{div}(t) = e^{-t}\mathrm{div}$ decays with t. $\mathrm{curl}(t) = e^{-t}\mathrm{curl}$.

(4) $a_{jk}H = x_j z_k - x_k z_j$. $\gamma_{jk}H = x_j z_k - x_k z_j$.

So H is rotationally invariant.

This completes our general discussion of moment theory. We now consider the orthogonal theory.

ORTHOGONAL THEORY

As for the case N=1 we look for generators $L(\underset{\sim}{z})$ such that the exponential $e^{\underset{\sim}{a}\cdot\overline{\underset{\sim}{C}}}$ has an orthogonal expansion generated by dual operators such that $\underset{\sim}{D}$ is mapped into a translation-invariant operator $\underset{\sim}{V}(\underset{\sim}{D})$. Let's call such an L Bernoulli after the general L of Theorem 2. We normalize $L(0) = \underset{\sim}{\nabla}L(0) = 0$, $\underset{\sim}{\nabla\nabla}L(0) = (\mathcal{E}_{jk})$, $\mathcal{E}_{jk}^2 = \delta_{jk}$.

Our basic equation $L'' = 1 + \alpha L' + \frac{\beta}{2}L'^2$ extends to:

$L(\underset{\sim}{z})$ is Bernoulli if it satisfies a system (of the type)

$$L_{jk} = \mathcal{E}_{jk} + \underset{\sim}{a}_{jk}\cdot\underset{\sim}{\nabla}L + B_{jk}\underset{\sim}{\nabla}L\cdot\underset{\sim}{\nabla}L$$

where $\mathcal{E}_{jk}^2 = \delta_{jk}$; a_{jk} is a vector, B_{jk} a matrix, for j,k fixed.

We proceed to clarify this equation.

<u>Proposition 59</u>: L satisfies such a system with some a_{jk}^{ℓ}, B_{jk}^{rs} symmetric in j,k.

<u>Proof</u>: We expand, for $z,w \in R^N$, $L(z+w) - L(z) - L(w)$ to second order around $z = 0$.

$$L(z+w) - L(z) - L(w) = z\cdot\underset{\sim}{\nabla}L(w) + \tfrac{1}{2}\Sigma\Sigma\, z_j z_k[L_{jk}(w) - \mathcal{E}_{jk}] + \ldots$$

For orthogonality we must have, for $\underset{\sim}{V}$: $R^N \to R^N$, $\underset{\sim}{D} \to \underset{\sim}{V}(\underset{\sim}{D})$,

$$\exp[L(z+w) - L(z) - L(w)] = \Sigma\, \frac{\underset{\sim}{V}(z)^{\underset{\sim}{n}}\underset{\sim}{V}(w)^{\underset{\sim}{n}}}{\underset{\sim}{n}!}k_{\underset{\sim}{n}} = \Sigma_{\underset{\sim}{n}} \Pi_j \frac{[V(z)_j V(w)_j]^{n_j}}{n_j!}\cdot k_{\underset{\sim}{n}}.$$

That is,

$$L(z+w) - L(z) - L(w) = \varphi[V_1(z)V_1(w),\ldots,V_N(z)V_N(w)].$$

We have directly $\underset{\sim}{V}(0) = 0$, $\varphi(0) = 0$.

Denote derivatives of φ with respect to its arguments by φ_j, φ_{jk}.

We see that

$$L_j(w) = \frac{\partial\varphi}{\partial z_j}\Big|_{z=0} = \Sigma\, \varphi_k(0)V_k(w)\,\frac{\partial V_k}{\partial z_j}(0) = \Sigma\, e_{jk}V_k(w)$$

$$L_{jk}(w) - \mathcal{E}_{jk} = \frac{\partial^2\varphi}{\partial z_j \partial z_k}\bigg|_{z=0} = \Sigma\Sigma\, p_{m\ell} V_\ell(w)\frac{\partial V_\ell}{\partial z_k}(0) V_m(w)\frac{\partial V_m}{\partial z_j}(0)$$

$$+ \Sigma\, p_m V_m(w)\frac{\partial^2 V_m}{\partial z_j \partial z_k}(0)$$

where $p_m = \varphi_m(0)$, $p_{m\ell} = \varphi_{m\ell}(0)$.

Assume $(\mathfrak{o}_{jk})$ is invertible with inverse (Θ_{jk}).

Then we have $\underset{\sim}{V} = \Theta\underset{\sim}{\nabla}L$.

This yields $L_{jk} = \mathcal{E}_{jk} + \sum_r a^r_{jk} L_r + \sum\sum_{rs} B^{rs}_{jk} L_r L_s$

with $a^r_{jk} = \sum_s p_s \Theta_{sr} \dfrac{\partial^2 V_s}{\partial z_j \partial z_k}(0)$ or $\underset{\sim}{a}_{jk} = \Theta^* \underset{\sim}{v}_{jk}$.

And $B_{jk} = \Theta^* P_{jk}\Theta$ where $P^{rs}_{jk} = p_{rs}\dfrac{\partial V_r}{\partial z_k}(0)\dfrac{\partial V_s}{\partial z_j}(0)$.

From $\underset{\sim}{V} = \Theta\underset{\sim}{\nabla}L$ we have $\dfrac{\partial V_r}{\partial z_k}(0) = \Theta_{rk}\mathcal{E}_{kk}$, $\dfrac{\partial^2 V_r}{\partial z_j \partial z_k}(0) = \sum_s \Theta_{rs} L_{sjk}(0)$.

Thus, $a^r_{jk} = \sum\sum_{sa} \Theta_{sr} p_s \Theta_{sa} L_{ajk}(0)$, $P^{rs}_{jk} = p_{rs}\Theta_{rk}\Theta_{sj}\mathcal{E}_{kk}\mathcal{E}_{jj}$.

Remarks:

1. We could write this

$$L'' = \mathcal{E} + a(L') + B(L',L')$$

where $L'' = \underset{\sim}{\nabla\nabla}L$, $L' = \underset{\sim}{\nabla}L$, $\mathcal{E}$ is a constant matrix, a is a linear mapping from $\mathbb{R}^N \to \mathbb{R}^{N^2}$ and B is a bilinear form $\mathbb{R}^N \times \mathbb{R}^N \to \mathbb{R}^{N^2}$.

2. The symmetry in j,k is immediate from the symmetry of L".
The symmetry of B^{rs}_{jk} in r and s follows directly from the symmetry of P^{rs}_{jk}.

3. We see that this is a Riccati PDE system in the vector L'.

4. $L' = \mathfrak{o}\underset{\sim}{V}$ with $\mathfrak{o}_{jk} = \varphi_k(0)\dfrac{\partial V_k}{\partial z_j}(0) = p_k \Theta_{kj}\mathcal{E}_{jj}$.

Let $P = \mathrm{diag}(p_k)$, i.e. $P_{ab} = \delta_{ab}p_b$.

Then $\mathfrak{o} = \mathcal{E}\Theta^* P$, i.e. $\Theta^{-1} = \mathcal{E}\Theta^* P$.

Thus Θ is orthogonal when $\mathcal{E} = P = I$.

5. The natural choice for a probabilistic process is $\mathcal{E}_{jk} = \delta_{jk}$ since this would be the diagonal matrix of var $w_j(1)$.
The processes $w_j(t)$ are, in any case, uncorrelated, i.e.
$\langle w_j(t) w_k(t)\rangle = t\mathcal{E}_{jk}$.

6. We will get symmetry properties of a^r_{jk}, B^{rs}_{jk} from the commutativity of the derivatives ∂_{z_j}; the usual "compatibility conditions" any system must satisfy.

We calculate
$$L_{jk\ell} = \Sigma\, a^r_{jk} L_{r\ell} + \Sigma\Sigma\, B^{rs}_{jk}(L_{r\ell}L_s + L_r L_{s\ell})$$
$$L_{jk\ell m} = \Sigma\, a^r_{jk} L_{r\ell m} + \Sigma\Sigma\, B^{rs}_{jk}(L_{r\ell m}L_s + L_{r\ell}L_{sm} + L_{rm}L_{s\ell} + L_r L_{s\ell m}).$$

At z=0 we see that
$$a^\ell_{jk} = \varepsilon_{\ell\ell} L_{jk\ell}(0) \quad \text{and so} \quad \varepsilon_{\ell\ell} a^\ell_{jk} = \varepsilon_{kk} a^k_{j\ell}.$$

From $\langle e^{\underset{\sim}{z}\cdot\underset{\sim}{w}(t)}\rangle = e^{tL(\underset{\sim}{z})}$ we have
$$\langle w_j(t) w_k(t) w_\ell(t)\rangle = tL_{jk\ell}(0) = t\varepsilon_{\ell\ell} a^\ell_{jk}.$$

Note that $|a^\ell_{jk}|$ is symmetric in all three indices.

From the fourth derivatives we have
$$L_{jk\ell m}(0) = \Sigma\, a^r_{jk} a^r_{\ell m}\varepsilon_{rr} + \Sigma\Sigma\, B^{rs}_{jk}(\varepsilon_{r\ell}\varepsilon_{sm} + \varepsilon_{rm}\varepsilon_{s\ell})$$
$$= \Sigma\, a^r_{jk} a^r_{\ell m}\varepsilon_{rr} + \varepsilon_{\ell\ell}\varepsilon_{mm}(B^{\ell m}_{jk} + B^{m\ell}_{jk})$$
$$= \Sigma\, a^r_{jk} a^r_{\ell m}\varepsilon_{rr} + 2\varepsilon_{\ell\ell}\varepsilon_{mm}B^{\ell m}_{jk}.$$

<u>Remark</u>: Observe that, since $L_{jk\ell m} = L_{\ell m jk}$, e.g. $\varepsilon_{\ell\ell}\varepsilon_{mm}B^{\ell m}_{jk} = \varepsilon_{jj}\varepsilon_{kk}B^{jk}_{\ell m}$.

Thus, given L, we can determine a,B from L''',L^{iv} and conversely from a,B we determine L through the basic equation. We thus state:

<u>Theorem 6</u> $L(\underset{\sim}{z})$ is a Bernoulli generator if and only if there are

(1) A matrix ε, $\varepsilon^2_{jk} = \delta_{jk}$,

(2) A matrix-vector $a^\ell_{jk} = \varepsilon_{\ell\ell} L_{jk\ell}(0)$,

(3) A matrix-matrix $B^{rs}_{jk} = \frac{1}{2}\varepsilon_{rr}\varepsilon_{ss}[L_{jkrs}(0) - \sum_p a^p_{jk} a^p_{rs}\varepsilon_{pp}]$,

such that
$$L_{jk} = \varepsilon_{jk} + \sum_\ell a^\ell_{jk} L_\ell + \sum\sum_{rs} B^{rs}_{jk} L_r L_s,$$
where L is normalized so that $L(0) = \underset{\sim}{\nabla}L(0) = 0$, $\underset{\approx}{\nabla\nabla}L(0) = \varepsilon$.

Furthermore, for any matrix Θ such that $\Theta^{-1} = \varepsilon\Theta^*$ we have the canonical operator $\underset{\sim}{V}(\underset{\sim}{D}) = \Theta\underset{\sim}{\nabla}L$.

<u>Proof</u>: The above discussion yields (1)-(3). The matrix $\mathcal{E}$ is essentially a choice of coordinates, which, as seen above, is essentially a choice of the correlation matrix $\langle w_j(1)w_k(1)\rangle$. We can eliminate (p_k) by noting that p_k can be absorbed into Θ, $\sqrt{p_j}\,\Theta_{jk} \to \Theta_{jk}$, equivalently $\sqrt{p_j}\,V_j \to V_j$.

However, p_{rs} becomes $\dfrac{p_{rs}}{\sqrt{p_r p_s}}$, as seen in the expression for P^{rs}_{jk}.

Given Θ, $P_{jk} = \Theta^* B_{jk}\Theta$;

so, as B_{jk} is chosen arbitrarily, as a parameter, P_{jk} is determined, i.e. the P_{jk} is absorbed into B_{jk}. Thus we can set the p's = 1.

EXAMPLES

We discuss two basic cases.

1. Brownian motion. This case is when $a=0$, $B=0$.

The system is $L_{jk} = \mathcal{E}_{jk}$ with solution $L = \frac{1}{2}\Sigma\, \mathcal{E}_{jj} z_j^2$, $V_j = \Sigma\, \Theta_{jk}\mathcal{E}_{kk} z_k$.

Thus, $\underset{\sim}{U} = \underset{\sim}{z} = \Theta^*\underset{\sim}{V}$. The generating function of the $J_{\underset{\sim}{n}}(\underset{\sim}{x},t)$ is thus,

$$\exp(\Sigma\Sigma\, x_j\Theta_{rj}v_r - \tfrac{1}{2}t\,\Sigma\, v_r^2) = \prod_r \exp[(\Theta x)_r v_r - \tfrac{1}{2}tv_r^2)$$

$$= \prod_{r=1}^{N} \Sigma \frac{v_r^{n_r}}{n_r!} H_{n_r}[(\Theta x)_r, t).$$

We check:

$$\Sigma\, \mathcal{E}_{jj} z_j^2 = \Sigma\, \mathcal{E}_{jj}(\Sigma\, \Theta_{rj}V_r)^2 = \Sigma\Sigma\Sigma\, \mathcal{E}_{jj}\Theta_{rj}\Theta_{sj}V_rV_s$$

$$= \sum_{rs}\sum V_rV_s \sum_{pj}\sum \Theta_{rp}\mathcal{E}_{pj}\Theta^*_{js} = \sum_{rs}\sum V_rV_s\delta_{rs}.$$

For $\Theta = \mathcal{E} = I$ we have $w_j(t)$ N independent 1-dimensional Brownian motions.

2. Poisson process. In this case $B=0$, a is arbitrary.

We consider the case $\mathcal{E}_{jk} = \delta_{jk}$, $N = 2$. We have the system

$$L_{zz} = 1 + aL_z + bL_w$$

$$L_{zw} = bL_z + eL_w$$

$$L_{ww} = 1 + eL_z + fL_w$$

The coefficients a^{ℓ}_{jk} are symmetric in j,k,ℓ, since $\mathcal{E}_{jk} = \delta_{jk}$.

Thus, $a^2_{11} = a^1_{12}$ and $a^1_{22} = a^2_{12}$ as indicated.

We consider a change of coordinates

$$z' = w-pz$$
$$w' = w-qz$$

Set $L(z,w) = F(w-pz) + G(w-qz)$.
Then we have

$$L_{ww} = F'' + G'' = 1 - epF' - eqG' + fF' + fG'.$$

Since F and G are functions of independent coordinates we have

$$F'' - (f-ep)F' = \text{constant} = -G'' + (f-eq)G' + 1.$$

In general we have an equation $y''-hy' = k^{-1}$ which has for solution, with

$$y(0) = y'(0) = 0, \quad y(s) = \frac{e^{hs}-1-hs}{kh^2} \equiv \frac{1}{k}E_h(s).$$

Thus, $\quad L(z,w) = \frac{1}{k}E_r(w-pz) + (1-\frac{1}{k})E_s(w-qz),$ where $r = f-ep$, $s = f-eq$.

We still have to satisfy the other two equations, for L_{zz}, L_{zw}.

<u>Proposition 60</u>: 1. For N=2 the generator of a Poisson process is of the form

$$L(z,w) = \frac{1}{k}E_r(w-pz) + \frac{1}{k'}E_s(w-qz)$$

with r,s arbitrary, $pq = -1$, $k = 1+p^2$, $k' = \frac{k}{k-1} = 1+q^2$;

$$E_t(y) \equiv \frac{e^{ty}-1-ty}{t^2}.$$

2. The general Poisson process $[x(t),y(t)] \in R^2$ satisfies

$$x(t) = p\nu_1(t) + q\nu_2(t)$$
$$y(t) = \nu_1(t) + \nu_2(t),$$

where $pq = -1$ and for r,s arbitrary, $k = 1+p^2$, $k' = 1+q^2$,

$$\nu_1(t) = r[N_1(\frac{t}{r^2k}) - \frac{t}{r^2k}], \quad \nu_2(t) = s[N_2(\frac{t}{s^2k'}) - \frac{t}{s^2k'}],$$

with $N_1(t)$, $N_2(t)$ independent standard Poisson processes $\in R^1$.

<u>Proof</u>: 1. It is easy to verify by substituting into the equations for L_{zz}, L_{zw}, L_{ww} that p and q are solutions to $x^2 + \frac{b-f}{e}x-1 = 0$;
$r = f-ep$, $s = f-eq$, $k = 1+p^2$, $k' = 1+q^2$, $k^{-1}+k'^{-1} = 1$.

There are 3 parameters b,f,e determining the three quantities $p+q$, $rs = f^2-f-e^2+b$ and $r+s = f+b$ so that r and s are arbitrary and p,q are subject only to the condition $pq = -1$.

2. We have, with $\nu_1(t)$, $\nu_2(t)$ as above,

$$\langle e^{zx+wy}\rangle = \exp[\tfrac{t}{k}E_r(w-pz)]\cdot\exp[\tfrac{t}{k'}E_s(w-qz)]$$
$$= \langle e^{(w-pz)\nu_1+(w-qz)\nu_2}\rangle.$$

Comparing coefficients of z and w in the exponents we have

$$x(t) = -p\nu_1-q\nu_2, \quad y(t) = \nu_1+\nu_2.$$

Since p,q are subject only to the condition $pq = -1$, we replace $p = -p'$, $q = -q'$.

It is clear that in outline the theory for N=1 extends in a fairly straightforward manner, but the terrain is much more rich and varied. Our theory gives immediate interesting extensions to $\mathbb{R}^N$ of the functions encountered for $\mathbb{R}^1$. We will discuss now some general features.

CANONICAL VARIABLES

The generating function $G(\underset{\sim}{x},t;\underset{\sim}{v})$, of the orthogonal polynomials $J_{\underset{\sim}{n}}(\underset{\sim}{x},t)$ is

$$G(\underset{\sim}{x},t;\underset{\sim}{v}) = \exp(\underset{\sim}{x}\cdot\underset{\sim}{U}-tM)$$

where $\underset{\sim}{U} = \underset{\sim}{V}^{-1}$ around 0 in a neighborhood of $\mathbb{C}^N$ and $M = L\circ\underset{\sim}{U}$.
The canonical gradient operator is $\underset{\sim}{V}(\underset{\sim}{D})$ which, up to choice of a matrix Θ satisfying $\Theta^{-1} = \mathcal{E}\Theta^*$, can be taken to be $\underset{\sim}{\nabla}L$.
The dual operator, corresponding to $\xi = xW$, satisfies

$$e^{\underset{\sim}{v}\cdot\underset{\sim}{\xi}}1 = G(\underset{\sim}{x},0;\underset{\sim}{v}) = e^{\underset{\sim}{x}\cdot\underset{\sim}{U}}$$

Applying $\frac{\partial}{\partial v_j}$ we have $\xi_j = \Sigma\, x_r \frac{\partial U_r}{\partial v_j}$. This implies

<u>Proposition 61</u>: 1. The canonical variables for $\underset{\sim}{w}(t) \in \mathbb{R}^N$ are $\underset{\sim}{V} = \Theta\underset{\sim}{\nabla}L$ and $\underset{\sim}{\xi} = \underset{\sim}{x}\cdot W$, where $WL''\Theta = I$, $\Theta^*\Theta = \mathcal{E}$.

2. $J_{\underset{\sim}{n}}(\underset{\sim}{x},t) = \underset{\sim}{\xi}(t)^{\underset{\sim}{n}}1$, where $\underset{\sim}{\xi}(t) = e^{-tL}\underset{\sim}{\xi}e^{tL} = \underset{\sim}{C}\cdot W$.

3. The number operator is $\underset{\sim}{\xi}(t)\cdot\underset{\sim}{V} = (\underset{\sim}{C}\cdot W)\cdot\underset{\sim}{V}$.

<u>Remark</u>: For a vector $\underset{\sim}{x}$ and matrix B, $(\underset{\sim}{x}\cdot B)_j = \Sigma\, x_r B_{rj}$, i.e. $\underset{\sim}{x}\cdot B = \underset{\sim}{x}^*B$.

<u>Proof</u>: 1. From $U[V(z)] = z$ we have, applying ∂_{z_k}, $\Sigma \frac{\partial U_r}{\partial v_j}\frac{\partial V_j}{\partial z_k} = \delta_{rk}$.

Choosing $V = \Theta\nabla L$, $\frac{\partial V_j}{\partial z_k}$ are elements of the matrix $\Theta L''$.

Thus, $\left(\frac{\partial U_r}{\partial v_j}\right) = (\Theta L'')^{-1}$.

2. follows directly from 1.

3. also follows directly.

<u>Remarks</u>:

1. In the following we will choose $\Theta = I$ so that

$$\Theta^*\Theta = I = \mathcal{E} = L''(0) \quad \text{and} \quad V = L'.$$

2. The number operator $(\bar{\underset{\sim}{C}}\cdot W)\cdot \underset{\sim}{V}$ is an extension to $\mathbb{R}^N$ of the confluent hypergeometric operator. We thus see that in general a confluent hypergeometric function of order n is any function homogeneous of degree n in the variable $\bar{\underset{\sim}{C}}\cdot W$.

3. We define the vacuum space to consist of those functions Ω such that
(1) Ω is an absolute vacuum function (for L),
(2) $V\Omega = 0$.
On this space the operators $\xi(t)$ and V are duals isomorphic to x,D as ordinary functions. E.g. $Vf(\xi)\Omega = [V,f(\xi)]\Omega = f'(\xi)\Omega$.

<u>Definition</u>: An operator <u>on the vacuum space</u> means that it applies to an operator function that in turn is applied to some fixed Ω in the vacuum space. E.g. On the vacuum space, $Vf(\xi) = f'(\xi)$.

We then have directly from Proposition 56,

<u>Proposition 62</u>: On the vacuum space,

$$(\bar{\underset{\sim}{C}}\cdot W)\cdot \underset{\sim}{V} = k\frac{d}{dk} + ty\frac{dM}{dy}, \quad \text{where } k = |\underset{\sim}{\xi}|, \quad y = |\underset{\sim}{v}|.$$

FURTHER REMARKS. RECURSION FORMULAS

The basic results for N=1 generalize readily. We have

$$p_t(\underset{\sim}{y}-\underset{\sim}{x}) = e^{tL}\delta(\underset{\sim}{x}-\underset{\sim}{y}) \quad \text{where} \quad \delta(\underset{\sim}{x}-\underset{\sim}{y}) = \prod_{j=1}^{N}\delta(x_j-y_j).$$

$$j_{\underset{\sim}{n}}(t) = \langle J_{\underset{\sim}{n}}^2\rangle. \quad \text{E.g.,}$$

<u>Proposition 63</u>: GRF

Let $f^* = \Sigma \frac{\underset{\sim}{x}^{\underset{\sim}{n}}}{\underset{\sim}{n}!} f_{\underset{\sim}{n}}$, where $f_{\underset{\sim}{n}} = \langle fJ_{\underset{\sim}{n}} \rangle$. Let $V^* = \underset{\sim}{V}(-\underset{\sim}{D})$.

Then

$$f(\underset{\sim}{x}) = \frac{1}{p_t(\underset{\sim}{x})} f^*(\underset{\sim}{V}^*) p_t(\underset{\sim}{x}).$$

The proof is as for $N = 1$.

We calculate the first few J's. We denote by e_j the vector with components δ_{jk}.

<u>Proposition 64</u>: $J_0 = 1, \quad J_{e_j} = x_j, \quad J_{e_j+e_k} = x_j x_k - \Sigma a^r_{jk} x_r - t\delta_{jk}.$

<u>Proof</u>: $J_0 = 1 = G(\underset{\sim}{x}, t; 0). \quad J_{e_j} = \Sigma x_r W_{rj} - tV_r W_{rj} 1 = \Sigma x_r \delta_{rj} = x_j.$

$$J_{e_j+e_k} = (\Sigma x_r W_{rk} - tV_r W_{rk}) x_j.$$

And

$$(\Sigma x_r W_{rk} - tV_r W_{rk}) x_j = \Sigma x_r x_j \delta_{rk} - t \Sigma V_{rj}(0) W_{rk}(0) + \Sigma x_r W_{rk,j}(0)$$

$$= x_j x_k - t\delta_{jk} - \Sigma a^r_{jk} x_r.$$

Where we check $W_{rk,j}(0) = -a^r_{jk}$ as follows.

From $\sum_s L_{rs} W_{sk} = \delta_{rk}$, we have $\Sigma L_{rsj} W_{sk} + L_{rs} W_{sk,j} = 0$;

at 0, $W_{sk}(0) = \delta_{sk}, L_{rs}(0) = \delta_{rs}$,

so that $W_{rk,j}(0) = -L_{rkj}(0) = -a^r_{jk}$.

The Expansion Theorem generalizes immediately. And consequently the Duality Theorem,

$\langle fJ_{\underset{\sim}{n}} \rangle = \frac{J_{\underset{\sim}{n}}}{\underset{\sim}{n}!} \langle \underset{\sim}{V}^{\underset{\sim}{n}} f \rangle$ is as for $N = 1$.

We can now proceed with the recurrence formula. We recall that $L_{jk\ell}(0) = a^{\ell}_{jk}$ and $L_{jk\ell m}(0) = \Sigma a^r_{jk} a^r_{\ell m} + 2B^{\ell m}_{jk}$.

<u>Proposition 65</u>: Recurrence Formula.

$$J_{n+e_j} = (x_j - \sum_r n_r a^r_{jr}) J_n - n_j [t + (n_j - 1) B^{jj}_{jj} + 2 \sum_{r \neq j} n_r B^{jr}_{jr}] J_{n-e_j}$$

$$- \sum_{p \neq q} n_p a^p_{qj} J_{n-e_p+e_q}.$$

Proof: The leading terms of $J_{\underset{\sim}{n}}$ are $[\Sigma\, x_r W_{rj}(0)]^{n_j}$, i.e. $x_j^{n_j}$.

Thus, $J_{\underset{\sim}{n}+e_j} - x_j J_{\underset{\sim}{n}}$ should be expressible in terms of $J_{\underset{\sim}{k}}$ where $|\underset{\sim}{k}| \leq |\underset{\sim}{n}|$. We have, for $|\underset{\sim}{k}| \leq |\underset{\sim}{n}|$,

$$\langle (J_{\underset{\sim}{n}+e_j} - x_j J_{\underset{\sim}{n}}) J_{\underset{\sim}{k}} \rangle = -\langle x_j J_{\underset{\sim}{k}} J_{\underset{\sim}{n}} \rangle .$$

In the following we use n,k to denote vectors $\underset{\sim}{n},\underset{\sim}{k}$ for clarity of notation. By the Duality Theorem,

$$\frac{n!}{j_n}\langle x_j J_k J_n \rangle = \langle V^n x_j J_k \rangle = \langle x_j V^n J_k \rangle + \langle \Sigma\, n_r V^{n-e_r} \frac{\partial V_r}{\partial z_j} J_k \rangle .$$

The first term is zero, as it is either zero since $k_r < n_r$ for some r, or it reduces to $\langle x_j \rangle = 0$.

Next we observe that

$$V^{n-e_r} J_k = 0 \text{ if } n_p > k_p,\ p \neq r, \text{ or } n_r > k_r + 1.$$

We thus have the conditions

$$1 + \Sigma\, k_b - n_b \geq \Sigma\, n_a - k_a \geq \Sigma\, k_b - n_b \geq 0$$

where a denotes an index such that $n_a > k_a$, b such that $n_b \leq k_b$. That is,

$$1 \geq (\Sigma\, n_a - k_a) - (\Sigma\, k_b - n_b) \geq 0, \quad 1 \geq |n| - |k| \geq 0.$$

Consider $V^{n-e_r} J_k \neq 0$. If $r \in \{a\}$, i.e. $n_r > k_r$, then $n_r = k_r + 1$ and the other indices are b's.

Since $0 \leq \Sigma\, k_b - n_b \leq 1 = \Sigma\, n_a - k_a$, either $k = n - e_r$ or $k = n - e_r + e_s$, $s \neq r$.

In the case $r \in \{b\}$, $n_p \leq k_p$ for $p \neq r$ as well, so $k = n$, i.e. all indices are b's. We have, then, three cases:

(1) $k = n$. $\frac{n!}{j_n}\langle x_j J_n J_n \rangle = n!\, \Sigma\, n_r \langle L_{rj} J_{e_r} \rangle = n!\, \Sigma\, n_r L_{rjr}(0) = n!\, \Sigma\, n_r a^r_{jr}$.

(2) $k = n - e_p$. $\frac{n!}{j_n}\langle x_j J_n J_k \rangle = \langle n_p V^{n-e_p} L_{pj} J_{n-e_p} \rangle = n! \langle L_{pj} 1 \rangle = n! \delta_{jp}$.

In case $p = j$, we have alternatively

$$\langle x_j J_n J_k \rangle = \frac{j_k}{k!} \langle V^{n-e_j} x_j J_n \rangle .$$

And

$$\langle V^k x_j J_n\rangle = \langle x_j V^{n-e_j} J_n\rangle + \langle \sum_{r\neq j} n_r V^{n-e_r-e_j} L_{rj} J_n\rangle + \langle (n_j-1) V^{n-2e_j} L_{jj} J_n\rangle$$

$$= (A) + (B) + (C).$$

(A) yields $\langle n! x_j J_{e_j}\rangle = n!\langle x_j^2\rangle = n!t.$

(B): $\langle \sum_{r\neq j} n_r n! L_{rj} J_{e_j+e_r}\rangle$. Using Proposition 64, for $j \neq r$,

$$\langle L_{rj} J_{e_j+e_r}\rangle = \langle L_{jr} x_j x_r - \Sigma\, a^s_{jr} L_{jrs}(0) - tL_{jr}\delta_{jr}\rangle$$

$$= L_{jrjr}(0) - \Sigma (a^s_{jr})^2 = 2B^{jr}_{jr}.$$

(C): $\langle V^{n-2e_j} L_{jj} J_n\rangle = \frac{n!}{2}\langle L_{jj} J_{2e_j}\rangle.$

$$\langle L_{jj} J_{2e_j}\rangle = \langle L_{jj}(x_j^2 - \Sigma\, a^s_{jj} x_s - t)\rangle = \langle x_j^2 L_{jj}(0)\rangle + L_{jjjj}(0) - \Sigma (a^s_{jj})^2 -$$

$$= 2B^{jj}_{jj}.$$

Combining the above we have, for $k = n-e_j$,

$$\langle x_j J_n J_k\rangle = j_n = j_{n-e_j} n_j [t + (n_j-1) B^{jj}_{jj} + 2 \sum_{r\neq j} n_r B^{jr}_{jr}].$$

(3) $k = n-e_p+e_q$, $p\neq q$.

$$\frac{n!}{j_n}\langle x_j J_n J_k\rangle = \langle n_p V^{n-e_p} L_{pj} J_{n-e_p+e_q}\rangle = n!(n_q+1)\langle L_{pj} J_{e_q}\rangle = n!(n_q+1) a^q_{pj}.$$

Alternatively we have

$$\frac{k!}{j_k}\langle x_j J_n J_k\rangle = \langle V^{n-e_p+e_q} x_j J_n\rangle = \langle x_j V^{n-e_p+e_q} J_n\rangle + \langle (n_q+1) V^{n-e_p} L_{qj} J_n\rangle$$

$$= (n_q+1) n! \langle L_{qj} J_{e_p}\rangle = (n_q+1) n! a^p_{qj}.$$

Thus, $\langle x_j J_n J_k\rangle = j_n (n_q+1) a^q_{pj} = j_k n_p a^p_{qj}.$

From the proof we get recursion formulas for j_n.

<u>Proposition 66</u>: 1. $j_n = j_{n-e_j}[t + (n_j-1) B^{jj}_{jj} + 2 \sum_{r\neq j} n_r B^{jr}_{jr}] = j_{n-e_j} n_j [t + (|n|-1)B]$,

where $B = B^{jj}_{jj}$, $1 \leq j \leq N$.

2. $j_n = \frac{n_p}{n_q+1} j_{n-e_p+e_q}$.

3. $B^{pp}_{pp} = B^{qq}_{qq} = 2B^{pq}_{pq}$.

4. For $j \neq k$ and any r, $B^{kk}_{jk} = B^{kr}_{jr} = B^{jk}_{kk} = 0$.

<u>Proof:</u> (1) follows from the (2) in the above proof and No. 3 below.

(2) follows from case $k = n-e_p+e_q$ using $a^q_{pj} = a^p_{qj}$.

(3) Apply No. 1 above to $n = ke_p + \ell e_q$, first for $e_j = e_q$, then for $e_j = e_p$:

$$j_{ke_p+\ell e_q} = j_{ke_p+(\ell-1)e_q} \ell[t+(\ell-1)B(q)+2kB^{pq}_{pq}]$$

$$= j_{(k-1)e_p+\ell e_q} k[t+(k-1)B(p)+2\ell B^{pq}_{pq}],$$

where $B(a) = B^{aa}_{aa}$.

Using No. 2 for $n = ke_p+(\ell-1)e_q$, we have

$$(\ell-1)B(q) - (k-1)B(p) = 2(\ell-k)B^{pq}_{pq}.$$

Setting $\ell = k$ we see that $B(q) = B(p)$. Thus, $B(p) = 2B^{pq}_{pq}$.

(4) In case (2) of the proof of Proposition 65 we have for $p \neq j$,

$\langle x_j J_n J_{n-e_p} \rangle = 0$.

Using the alternative calculation we have the corresponding

(A): $n! \langle x_j x_p \rangle = 0$,

(B): $L_{jrpr}(0) - \Sigma a^s_{pr} L_{jrs}(0) = 2B^{pr}_{jr}, \quad p \neq r$.

(C): $\langle (n_p-1) V^{n-2e_p} L_{jp} J_n \rangle = \frac{(n_p-1)n!}{2} \langle L_{jp} J_{2e_p} \rangle$.

And $\langle L_{jp} J_{2e_p} \rangle = \langle L_{jp}(x^2_p - \Sigma a^s_{pp} x_s - t) \rangle = L_{jppp}(0) - \Sigma a^s_{pp} L_{jps}(0)$

$$= 2B^{pp}_{jp}.$$

Thus, $2 \sum_{r \neq p} n_r B^{pr}_{jr} + (n_p-1) B^{pp}_{jp} = 0$.

Since this holds for arbitrary n, the B's must be zero.

From the symmetry of $L_{jk\ell m}$ we have

$$|B^{pp}_{jp}| = |B^{jp}_{pp}| = 0.$$

Remarks: 1. From No. 2 of Proposition 66 we can express the recurrence formula in the form:

$$J_{n+e_j} = x_j J_n - \sum_{rs}\sum n_r a^r_{js} J_{n-e_r+e_s} - n_j[t+(|n|-1)B]J_{n-e_j}.$$

2. We have

$$j_{\underset{\sim}{n}} = \underset{\sim}{n}!t(t+B)\ldots[t+(n_1-1)B](t+n_1B)(t+n_1B)\ldots[t+(n_1+n_2-1)B]\ldots$$

$$\ldots[t+(|\underset{\sim}{n}|-1)B] = B^{|\underset{\sim}{n}|}\underset{\sim}{n}!\left(\tfrac{t}{B}\right)_{(|\underset{\sim}{n}|)}.$$

SOME REMARKS ON THE GENERAL BERNOULLI GENERATOR

We see that in the general case we can organize B^{rs}_{jk} into a special large matrix with four blocks. First, order the equations: $L_{11}, L_{22}, \ldots, L_{NN}, L_{12}, L_{13}, \ldots, L_{1N}, L_{23}, L_{24}, \ldots, L_{2N}, \ldots, L_{N-1,N}$.
Then the upper $N \times N$ block is B^{rr}_{jj} and the upper right and lower left blocks consist of terms relating L_{jj} and L_rL_s, the lower right block relating L_{jk}, L_rL_s. The diagonal is $B^{jj}_{jj} = 2B^{pq}_{pq} = B$, constant. And $|B^{rs}_{jk}|$ is a symmetric matrix. Regardless of the zero terms of B^{rs}_{jk} (Proposition 66) we will see that generally there is a linear transformation that separates the variables.

We set $L(\underset{\sim}{z}) = \sum_1^N \varphi_j(Z_j)$ where $\underset{\sim}{Z} = R\underset{\sim}{z}$.

Then
$$\sum_s R_{sj}R_{sk}\varphi''_s = \mathcal{E}_{jk} + \sum_{s\ell}\sum a^{\ell}_{jk}R_{s\ell}\varphi'_s + \sum\sum\Big(\sum_{ab}\sum_{pq} B^{pq}_{jk}R_{ap}R_{bq}\Big)\varphi'_a\varphi'_b.$$

There are $N + \frac{N(N-1)}{2} = \frac{N(N+1)}{2}$ equations and $\frac{N(N-1)}{2}$ pairs $\varphi'_a\varphi'_b$, $a \neq b$ forming the quadratic cross-terms. Thus, there is a linear combination of, say, the first $\frac{N(N-1)}{2}+1$ equations that eliminates these cross-terms. The resulting equation is a sum of equations for the independent functions φ_j. The constants α_j, β_j for each generator $\varphi_j(Z_j)$ are, of course, not independent in general, but we have

$$\langle e^{\underset{\sim}{z}\cdot \underset{\sim}{w}(t)}\rangle = e^{tL(\underset{\sim}{z})} = e^{t\sum\varphi_j(Z_j)} = \langle e^{\underset{\sim}{Z}\cdot\underset{\sim}{x}(t)}\rangle$$

where $\underset{\sim}{x}(t)$ has N independent components $x_j(t)$ with generators $\varphi_j(D_j)$.
From $\underset{\sim}{z}\cdot\underset{\sim}{w}(t) = R\underset{\sim}{z}\cdot\underset{\sim}{x}$ we conclude that $\underset{\sim}{w}(t) = R^*\underset{\sim}{x}(t)$.
Thus, a general Bernoulli process on R^N is a linear transformation of a process with independent components; the parameters α_j, β_j determining the independent generators are, however, not independent.

We conclude our discussion with remarks on the Lagrangians for multidimensional processes.

LAGRANGIANS

The Lagrangians can be found for generators for processes on $\mathbb{R}^N$. We have

$$H[R'(\alpha)] = \Sigma \frac{\partial R}{\partial \alpha_j} \alpha_j - R(\underset{\sim}{\alpha})$$

where R is the Lagrangian, $\alpha_j = \dot{x}_j$, $R'(\alpha) = \nabla_{\underset{\sim}{\alpha}} R$. Applying ∂_{α_k} we have

$$\Sigma H_s R_{sk} = \Sigma R_{jk}\alpha_j + R_k - R_k, \quad \text{i.e.} \quad \underset{\sim}{H}' \cdot R'' = \underset{\sim}{\alpha} \cdot R''.$$

Assuming R" is nonsingular, $H'[R'(\underset{\sim}{\alpha})] = \underset{\sim}{\alpha}$.

For $H = L$, $H' = V$, $U = V^{-1}$, $R'(\alpha) = \underset{\sim}{U}(\underset{\sim}{\alpha})$.

Thus $$H[R'(\alpha)] = L[\underset{\sim}{U}(\underset{\sim}{\alpha})] = M(\underset{\sim}{\alpha}) = \underset{\sim}{\alpha} \cdot \underset{\sim}{U}(\alpha) - R(\underset{\sim}{\alpha}).$$

Thus, $R(\underset{\sim}{\alpha}) = \underset{\sim}{\alpha} \cdot \underset{\sim}{U}(\underset{\sim}{\alpha}) - M(\underset{\sim}{\alpha})$. And $\underset{\sim}{V}(\underset{\sim}{z})$ is the velocity as a function of the momentum — $\underset{\sim}{z}$. We remark that for the case of independent components, the total Lagrangian is the sum of the individual ones, thus corresponding physically to non-interacting particles.

This concludes our discussion of the basic theory for $\mathbb{R}^N$. Clearly there is much to clarify and especially interesting would be an exploration of the function theory associated with the various g's and K_t's generalizing, e.g. Bessel functions, for $\mathbb{R}^N$ and $\mathbb{C}^N$.

CHAPTER IX. HAMILTONIAN PROCESSES AND STOCHASTIC PROCESSES

In this chapter we study more thoroughly the relationship between the process $w(t)$ and the operator path $x(t)$ generated by $H = L$. We consider general $H(x,z)$ such that $H1 = 0$. We assume there exists a "Markov process" generated by H even though the density $e^{tH}\delta(x-y)$ may not be a positive measure.

ANALYTICAL MARTINGALES

Our first topic is to consider the physical paths corresponding to martingales. In the theory of Markov processes we can define a process by the condition that

$$f[w(t),t] - \int_0^t (\frac{\partial}{\partial s}+H)f[w(s),s]ds$$

is a martingale for every f in some sufficiently broad class of functions (e.g. $f \in S$ or C_0^∞). We will consider this in the context of our theory.

For a (general) Markov process generated by H, we have

$$\langle f[w(t)]\rangle_x = f[x(t)]1$$

since $$\langle e^{aw(t)}\rangle_x = e^{tH}e^{ax} = e^{tH}e^{ax}e^{-tH}1 = e^{ax(t)}1.$$

By the Markov property, for $\mathcal{F}_s = \sigma\{w(s'),\ 0 \le s' \le s\}$,

$$E(e^{aw(t)}\|\mathcal{F}_s) = E[e^{aw(t)}\|w(s)] = e^{(t-s)H(x,D)}e^{ax}\Big|_{x=w(s)}.$$

Another way of seeing this is $f[w(t)] = e^{w(t)D}f(0)$ so that

$$E[f(w(t))\|w(s)] = E[e^{w(t)D-w(s)D}\|w(s)]\cdot f[w(s)]$$
$$= e^{(t-s)H[w(s),D]}f[w(s)],$$

using time-homogeneity. We can express this by

$$E[f(w(t))\|w(s)] = f[x(t-s)]1\Big|_{x=w(s)}.$$

Proposition 67: If $u(x,t)$ satisfies $\frac{\partial u}{\partial t} + Hu = 0$, then $u[w(t),t]$ is a martingale.

Proof: $$E[u(w(t),t) - u(w(s),s)\|\mathcal{F}_s] = E[e^{-tH}u(w(t),0)\|\mathcal{F}_s] - e^{-sH}u(w(s),0).$$

$$E[e^{-tH}u(w(t),0)\|\mathcal{F}_s] = e^{(t-s)H}e^{-tH}u(w(s),0) = e^{-sH}u(w(s),0).$$

Note that, e.g., $e^{-tH}u(w(t),0)$ means $e^{-tH(x,D)}u(x,0)\Big|_{x=w(t)}$.

Remark: An even simpler way to see this in the case $H = L(z)$ is to consider the basic exponential martingale $e^{aw(t)-tL(a)}$. Then

$$e^{w(t)D-tL(D)}u(0,0) = e^{-tL(D)}u(x,0)\Big|_{x=w(t)}.$$

Thus for $u(x,t) = e^{-tL}u(x,0)$,

$$u[w(t),t] = e^{w(t)D-tL(D)}u(0,0) \quad \text{is a martingale.}$$

In this sense, then, the exponential martingale "generates" the analytical martingales.

We have seen that if $\frac{\partial u}{\partial t} + Hu = 0$, then $u[x(t),t]1$ is independent of t. This agrees since

$$\langle u[w(t),t]\rangle = u(x,0) = u[x(t),t]1.$$

The martingale on the physical space is thus the same as a "conserved quantity." In terms of operators we have seen that harmonic operators generate the symmetry groups of the flow. Thus, martingales correspond to the symmetries of a process.

Consider $f[x(t),t] = e^{tH}f(x,t)e^{-tH}$. Then $\frac{df}{dt}[x(t),t] = (\frac{\partial f}{\partial t} + [H,f])[x(t),t]$.

We thus have

$$f[x(t),t] - f[x(s),s] = \int_s^t (\frac{\partial f}{\partial a} + [H,f])[x(a),a]da.$$

Now consider $E[f(w(t),t) - f(w(s),s)\|\mathcal{F}_s]$. Since we are assuming time homogeneity this is the same as

$$\langle f[w(t-s),s+t-s] - f(x,s)\rangle_{x=w(s)} = f[x(t-s),s+t-s]1 - f(x,s)\Big|_{x=w(s)}.$$

From the above equation for paths $x(t)$ we have

$$\begin{aligned} f[x(t-s),s+t-s]1 - f(x,s) &= \int_0^{t-s} (\frac{\partial f}{\partial a} + [H,f])[x(a),a+s]1\, da \\ &= \int_0^{t-s} (\frac{\partial f}{\partial a} + Hf)[x(a),a+s]1\, da \quad \text{since} \quad H1 = 0. \end{aligned}$$

We thus get

$$\begin{aligned} E(f[w(t),t] - f[w(s),s]\|\mathcal{F}_s) &= \int_0^{t-s} \langle (\frac{\partial f}{\partial a} + Hf)[w(a),a+s]\rangle_{w(s)} da \\ &= E(\int_s^t [\frac{\partial f}{\partial a} + Hf][w(a),a]da\|\mathcal{F}_s). \end{aligned}$$

That is,

<u>Proposition</u> <u>68</u>: $f[w(t),t] - \int_0^t (\frac{\partial f}{\partial s} + Hf)[w(s),s]ds$ is a martingale.

Correspondingly, we see from the formula above that in the $x(t)$-space

$$f[x(t),t]1 - \int_0^t (\frac{\partial f}{\partial a} + Hf)[x(a),a]1\, da$$

is constant in time, as it equals $f[x(0),0] = f(x,0)$. Observe that we could apply $f[x(t)]$ to any vacuum function Ω and still have properties analogous to expectation. The only property of 1 that is really necessary is that $H1 = 0$.

FUNCTIONALS OF PROCESSES

In quantum field theory the processes considered take values in S^* or another space of distributions. Here we consider the relation between probabilistic paths $w(t) \in \mathbb{R}$ and $x(t) \in$ operator space.

We consider $\langle f[w(t)]\rangle_x = f[x(t)]1$ with $f = \chi_B$ for a Borel set B.

So we have

$$P_x[w(t)\in B] = \chi_B[x(t)]1.$$

Similarly,

$$P_x[w(t_1)\in B_1, w(t_2)\in B_2] = \int \chi_{B_1}(y)p_{t_1}(x,dy) \int \chi_{B_2}(z)p_{t_2-t_1}(y,dz)$$

$$= e^{t_1 H}\chi_{B_1} e^{(t_2-t_1)H}\chi_{B_2} e^{-t_2 H} 1$$

$$= \chi_{B_1}[x(t_1)]\chi_{B_2}[x(t_2)]1.$$

By induction we see that for $t_1 < t_2 < \ldots < t_n$,

$$P_x[w(t_j)\in B_j, 1\leq j\leq n] = \prod_{j=1}^{n} \chi_{B_j}[x(t_j)]1,$$

where the largest time is on the right and then backwards to t_1.

We denote by Φ_* a functional that is time-ordered, i.e. it is a limit of functionals depending on only finitely many time points each approximation being ordered starting with the largest time on the right.

We have, then,

<u>Theorem 7</u> Let Φ be in $L^1(d\mu)$ where $d\mu$ is the measure on paths $w(t)$. Then

$$\langle\Phi(w)\rangle_x = \Phi_*[x(t)]1.$$

<u>Proof:</u> Since Φ is measurable it can be approximated by finite-dimensional functionals and convergence to Φ by the approximations has a limit since $\Phi \in L^1$.

<u>Examples:</u> 1. A simple example is $\langle\int_0^t w(s)ds\rangle_x = \int_0^t x(s)ds\ 1$.

For $H = L(z)$, $x(s) = x + sL'(z)$ and

$$\int_0^t x(s)1\,ds = \int_0^t (x+s\mu)ds = xt + \frac{\mu t^2}{2}, \quad \mu = \langle w(1)\rangle_0 = L'(0).$$

2. $\langle [\int_0^t w(s)ds]^2\rangle_0 = \langle \int_0^t\int_0^t w(s)w(s')dsds'\rangle_0 = 2\iint_{s<s'} x(s)x(s')1dsds'\Big|_{x=0}.$

For $H = L(z)$,

$$L'(0) = \mu, \quad L''(0) = \sigma^2,$$

$$\langle w(s)w(s')\rangle = \langle [w(s') - w(s)]w(s)\rangle + \langle w(s)^2\rangle = (s'-s)s\mu^2 + s\sigma^2 + s^2\mu^2$$
$$= s's\mu^2 + s\sigma^2.$$

And $\quad x(s)x(s')1 = (x+sL')(x+s'L')1 = (x+sL')(x+s'\mu)$

$$= x^2 + xs'\mu + sx\mu + sL''(0) + ss'\mu^2$$

which reduces accordingly at $x=0$.

3. We consider the example near the end of Chapter III.

$W(t)$ is generated by $L'(z)$.

Then $\quad \langle \exp[\int_0^a W(st)ds]\rangle_0 = e^{tL(a)}.$

We approximate the integral by a sum

$$\Sigma W(s_j t)\Delta s_j, \quad 0 = s_0 < s_1 < s_2 < \dots < s_{n-1} < s_n = a.$$

Then $\quad \Phi_*[x(t)]1 = \lim_{n\to\infty} \prod_{j=1}^{n}{}_* e^{x(s_j t)\Delta s_j}1.$

Using $e^{ax(t)} = e^{tH}e^{ax}e^{-tH}$, $\quad H = L'$, $\;$ we have

$$\prod_{j=1}^{n}{}_* e^{x(s_j t)\Delta s_j}1 =$$

$$e^{s_1 tH}e^{x\Delta s_1}e^{-s_1 tH}e^{s_2 tH}e^{x\Delta s_2}e^{-s_2 tH}e^{s_3 tH}\dots e^{\Delta s_n tH}e^{x\Delta s_n}e^{-atH}1$$

$$= \prod_{j=1}^{n}{}_* e^{\Delta s_j tH}e^{x\Delta s_j}$$

$$= e^{ax}e^{t\sum_0^{n-1} H(a-s_{n-j-1})\Delta s_{n-j}}$$

$$\xrightarrow[n\to\infty]{} e^{ax+t\int_0^a H(a-s)ds} = e^{ax+tL(a)} \quad \text{since } H = L'.$$

This example generalizes to a sketch of a proof of the Feynman-Kac formula which we proceed to discuss next.

THE FEYNMAN-KAC FORMULA

There are three formulations of interest of the Feynman-Kac formula in addition to the standard one. We have

<u>Theorem 8</u> Formulations of FK, the Feynman-Kac formula.

Let $q(x)$ be a potential function such that $q(iz)$ generates a process $k(t)$. Let $w(t)$ be a process with generator $L(z)$. Then the solution to $\frac{\partial u}{\partial t} = L(D)u + q(x)u$, $u(x,0) = f(x)$ can be given by any of four equivalent expressions:

1. $u = \langle f[w(t)]e^{\int_0^t q[w(s)]ds}\rangle_x$

2. $u = \langle e^{-ik(t)x} e^{\int_0^t L[D+i(k(s)-k(t))]ds}\rangle f(x)$

3. $u(a,t) = \langle e^{\int_0^t q[a+w^*(s,D)]ds}\rangle e^{tL}f(a)$,

   where $w^*(t,z)$ is generated by the <u>dual generator</u> $L^*(D) = L(D+z)-L(z)$.

4. $u = (e^{\int_0^t q[x(s)]ds})_* e^{tL}f(x)$.

<u>Remark</u>: Formulations 1 and 4 generalize directly to generators $H(x,z)$, $H1 = 0$.

<u>Proof</u>: (1) is the standard formulation.

(2) Let $v = \hat{u} = \frac{1}{2\pi}\int e^{-i\xi x}u(x,t)dx$. Then v satisfies

$$\frac{\partial v}{\partial t} = L(i\xi)v + q(i\partial_\xi)v, \quad v(\xi,0) = \hat{f}.$$

By No. 1, $v = \langle \hat{f}[\xi+k(t)]e^{\int_0^t L[i\xi+ik(s)]ds}\rangle$.

Thus, $u = \int e^{i\xi x}v(\xi,t) = \langle \int e^{i\xi x}\hat{f}[\xi+k(t)]e^{\int_0^t L[i\xi+ik(s)]ds}\rangle$

$$= \langle e^{-ixk(t)} e^{\int_0^t L[D+i(k(s)-k(t))]ds}\rangle f(x).$$

(3) Consider u such that $u(x,0) = e^{zx}$. Now set $h(x,t) = e^{-zx}u(x,t)$. Then h satisfies

$$\frac{\partial h}{\partial t} = e^{-zx}\frac{\partial u}{\partial t} = e^{-zx}Lu + e^{-zx}qu = [L(D+z)-L(z)]h + [L(z)+q(x)]h,$$

since $L(D+z)e^{-zx} = e^{-zx}L(D)$ by EL.

By No. 1, $h = \langle e^{\int_0^t [L(z)+q(x+w^*(s,z))ds]}\rangle = \langle e^{\int_0^t q[x+w^*(s,z)]ds}\rangle e^{tL(z)}$.

We thus have

$$u = e^{zx}h = \langle e^{\int_0^t q[x+w^*(s,z)]ds}\rangle e^{tL(z)}e^{zx}.$$

Substitute $z \to D = \partial_x$, $x \to a$. Then the operator $u_D(a,t)$ satisfies

$$\frac{\partial u_D}{\partial t} = L(\partial_a)u_D + q(a)u_D, \quad u_D(a,0) = e^{aD}.$$

Applying u_D to $f(0)$ yields $u(a,t)$, satisfying

$$\frac{\partial u}{\partial t} = L(\partial_a)u + q(a)u, \quad u(a,0) = f(a).$$

(4) The discussion of No. 4 includes our sketch of a proof of No. 1. From No. 1 we have

$$\langle e^{ax+aw(t)}e^{\int_0^t q[x+w(s)]ds}\rangle = e^{t(L+q)}e^{ax}.$$

By Theorem 7,

$$e^{t(L+q)}e^{ax} = e_*^{\int_0^t q[x(s)]ds}e^{ax(t)}1 = e_*^{\int_0^t q[x(s)]ds}e^{tL}e^{ax}.$$

By Proposition 1, No. 4 follows. Note that we could have used a similar approach for No. 3.

As in Example 3 of the previous section we see that

$e_*^{\int_0^t q[x(s)]ds}e^{tL}e^{ax}$ is approximated by

$$U_n e^{ax} = e^{s_1 L}e^{q(x)\Delta s_1}e^{\Delta s_2 L}e^{q(x)\Delta s_2}\ldots e^{(t-s_{n-1})L}e^{q(x)\Delta s_n}e^{-tL}e^{tL}e^{ax}$$

$$= \prod_{j=1}^{n-1}{}_* e^{\Delta s_j L}e^{q(x)\Delta s_j}e^{(t-s_{n-1})L}e^{q(x)(t-s_{n-1})}e^{ax}.$$

Applying $\frac{\partial}{\partial t}$ we have

$$\left(\prod_{1}^{n-1}{}_*\right)[e^{(t-s_{n-1})L}Le^{q(x)(t-s_{n-1})}e^{ax} + e^{(t-s_{n-1})L}e^{q(x)(t-s_{n-1})}q(x)e^{ax}].$$

The term $Le^{q(x)\Delta s_n} = e^{q(x)\Delta s_n}L + e^{q(x)\Delta s_n}q'(x)\Delta s_n L' + \ldots$, and thus $\frac{\partial U_n}{\partial t}$ converges to $U_n(L+q)$ as $\text{mesh}\{s_j\} \to 0$.

Thus $e_*^{\int_0^t q[x(s)]ds}e^{tL}e^{ax}$ satisfies $\frac{\partial u}{\partial t} = (L+q)u$, $u(x,0) = e^{ax}$.

And by Theorem 7, No. 1 follows.

Examples: 1. For Brownian motion, $L = D^2/2$, $L^* = D^2/2 + zD$.

Thus, $w^*(t) = b(t) + zt$.

And by No. 3,

$$\langle f[x+b(t)]e^{\int_0^t q[x+b(s)]ds}\rangle = \langle e^{\int_0^t q[a+b(s)+sD]}\rangle e^{tL} f(a)\Big|_{a=x}.$$

Setting $q = vx$, $f(x) = x^n$ yields

$$\langle [x+b(t)]^n e^{v\int_0^t b(s)ds}\rangle = \langle e^{v\int_0^t b(s)ds}\rangle e^{\frac{1}{2}vt^2 D} H_n(x,-t).$$

By our result on $\int_0^t w(s)ds$, we have $L = z^3/6$ for $L' = z^2/2$, and so

$$\langle [x+b(t)]^n e^{v\int_0^t b(s)ds}\rangle = e^{v^2t^3/6} H_n(x+\frac{vt^2}{2},-t).$$

2. For the Poisson process, $N(t), L = e^D - 1$, $L^* = e^z(e^D-1)$, $w^* = N(te^z)$.

By No. 3,

$$\langle f[x+N(t)]e^{\int_0^t q[x+N(s)]ds}\rangle = \langle f[x+N(t)]e^{\sum_0^{N(t)-1} q(x+j)\tau_j q_{N(t)}}\rangle$$

$$= \langle e^{\int_0^t q[a+N(se^D)]ds}\rangle e^{tL} f(a)\Big|_{a=x},$$

where τ_j are independent, exponentially distributed waiting times between the jumps of $N(t)$; $q_N = q(x+N)(t-T_N)$, $T_n = \tau_0 + \dots + \tau_{N-1}$.

Remarks: Thus we see that $\langle e^{\int_0^t q[a+w^*(s,D)]ds}\rangle = e^{tQ}e^{-tL}$, where $Q = L(D) + q(x)$.

And $\lim_{t\to\infty} e^{tQ}e^{-tL} = \langle e^{\int_0^\infty q[a+w^*(s,D)]ds}\rangle$ is analogous to a wave operator for potential scattering. E.g. in the Poisson case we have

$$\langle e^{\int_0^t q[a+N(se^D)]ds}\rangle f = \langle e^{\int_0^{te^D} q[a+N(s)]dse^D}\rangle f.$$

And so $\lim_{t\to\infty} e^{tQ}e^{-tL} f = \langle e^{\int_0^\infty q[a+N(s)]dse^D}\rangle f = \langle e^{e^D\sum_0^\infty q(a+j)\tau_j}\rangle f$

$$= \prod_0^\infty \frac{1}{1-q(a+j)e^D} f(x)\Big|_{x=a}.$$

In the x-space we have

$$e_*^{\int_0^t q[x(s)]ds} = e^{tQ}e^{-tL}.$$

3. Consider $\frac{\partial u}{\partial t} = \frac{1}{3}u''' - \frac{x^2}{2}u, \quad u(x,0) = f(x).$

The process generated by $q(i\partial_\xi) = \frac{1}{2}\partial_\xi^2$ is Brownian motion.

From No. 2, for the Airy process $a(t)$,

$$\langle f[x+a(t)]e^{-\frac{1}{2}\int_0^t [x+a(s)]^2 ds}\rangle = \langle e^{-ixb(t)}e^{\frac{1}{3}\int_0^t [D+ib(s)-ib(t)]^3 ds}\rangle f(x).$$

The dual operator $L^* = \frac{D^3}{3} + zD^2 + z^2 D$ so that

$w^*(t) = a(t) + zb(2t) + z^2 t$, where $a(t)$ and $b(t)$ are independent.

By No. 3, for $Q = D^3/3 - x^2/2$,

$$e^{tQ}e^{-tD^3/3} = \langle e^{-\frac{1}{2}\int_0^t [y+a(s)+b(2s)D+sD^2]^2 ds}\rangle\Big|_{y=x}$$

$$= \langle e^{-ixb(t)}e^{\frac{1}{3}\int_0^t [D+ib(s)-ib(t)]^3 ds}\rangle e^{-tD^3/3}.$$

We also remark that as seen in Example 1 above,

$$\langle e^{v\int_0^t b(s)ds}\rangle = e^{v^2 t^3/6},$$

that is, the Airy process is essentially a space-time dual of $\int_0^t b(s)ds$.

Some simple applications are to the exponential martingales.

Consider $e^{aw(t)-tL(a)} = e^{a[x+w(t)]-\int_0^t L(a)ds}e^{-ax}$.

By FK, $\langle e^{ax+aw(t)-tL(a)}\rangle$ satisfies $\frac{\partial u}{\partial t} = L(D)u - L(a)u, \quad u(x,0) = e^{ax}$.

The solution to this equation is clearly e^{ax}. This is the same as $\langle e^{aw(t)-tL(a)}\rangle = 1$.

Similarly for $H(x,z)$ we see that $e^{aw(t)-\int_0^t H[w(s),a]ds}$ is the exponential martingale.

By FK, $\langle e^{aw(t)-\int_0^t H[w(s),a]ds}\rangle_x$ satisfies $\frac{\partial u}{\partial t} = H(x,D)u - H(x,a)u, \quad u(x,0) = e^{ax}$,

which again clearly has the solution e^{ax} and thus

$$\langle e^{aw(t)-\int_0^t H[w(s),a]ds}\rangle_x = e^{ax}.$$

We have further

$$\langle e^{aw(t)-\int_0^t H[w(s),a]ds}\rangle_x = e_*^{-\int_0^t H[x(s),a]ds} e^{ax(t)}{}_1$$

$$= e_*^{-\int_0^t H[x(s),a]ds} e^{tH} e^{ax} = e^{ax}.$$

So $$e^{tH}e^{ax} = e_*^{\int_0^t H[x(s),a]} e^{ax}.$$

As in the proof of Theorem 8 we thus have

$$e^{tH(x,D)}f(x) = e_*^{\int_0^t H[y(s),D]} f(x)\Big|_{\substack{y(s)=x(s)\\ 0\le s\le t}}.$$

For a stochastic integral the corresponding martingale is

$$e^{\int_0^t a(s)dw(s)-\int_0^t H[w(s),a(s)]ds}.$$

This expression, for $a(s) = z(s)$ physical momentum and $w(s) = x(s)$, physical path (not operator path), is

$$e^{\int_0^t zdx-\int_0^t H[x(s),z(s)]ds} = e^{\int_0^t Rds}$$

where R is the Lagrangian, that is, this is exactly the exponential of the action. So Feynman's principle, in the "probabilistic limit," that the density of the measure generated by H on physical paths should be $e^{-\int Rdt}$ is somehow a transmutation of the fact that the expression $e^{\int Rdt}$, when z is replaced by an "arbitrary" function a(t) and x(t) replaced by w(t), is a martingale on the related probability space.

We have then the principle of correspondence between probabilistic processes and quantum (Hamiltonian) processes. Take the limit "$\frac{i}{\hbar} = -1$," i.e. $t' = i\hbar t$. Then determine a measure on (physical) paths that yields a probabilistic process by the condition that the exponential $e^{\int Rdt}$ of the action is a martingale under an arbitrary replacement of the momentum by any of a sufficiently broad class of functions.

CHAPTER X. CONCLUDING REMARKS

We recall some topics mentioned that need much elaboration and will be quite interesting to explore.

1. Technical questions concerning Chapter II require the theories of pseudo-differential operators and spectral synthesis (i.e. in more general settings can one use, e.g., e^{ax}, to generate everything).

2. Stochastic features of processes generated by D^n need to be studied, e.g. for the Airy process. The FK formula can be used to find, e.g. the distribution of $\sup_{0 \leq s \leq t} a(s)$.

3. From Chapter IV the question is how to find the V operator for (a) general measures that yield orthogonal polynomials, or (b) corresponding to a given $H(x,z)$. In case (b), the Hamilton-Jacobi theory should provide the "natural variables."

4. The results and ideas of Chapters IV, V and VIII, are clearly of wide application in the theory of special functions. E.g. are there general addition theorems for the functions that for particular cases are Bessel functions? For $\mathbb{R}^N$ we have canonical constructions generalizing the special functions on $\mathbb{R}^1$ and $\mathbb{C}$.

5. Chapter VI requires theorems on weak convergence of the appropriate measures.

6. Chapter VII shows the basic elements from which the "smooth" cases are constructed. There should be a version of the theory for $\mathbb{R}^N$, that is, with a $N \times n$'ple array $x^1_{jn} \ldots x^N_{jn}$ as the nth approximation to $dx^1 \ldots dx^N$. The "β" problem of constructing the J_n's of Chapter IV may require more complicated theory of representation of semimartingales.

7. We still need the complete solution to the equation for Bernoulli generators on $\mathbb{R}^N$.

8. The theory of correspondence between physical and probabilistic processes is still just getting going, even in quantum field theory. There the discrete approximation (Chapter VII here) is analyzed by statistical mechanics. There should also be a relativistic probability theory by mapping back from relativistic quantum theory.

9. The exact meaning of the "probabilistic limit" $t \to i\hbar t$ is still unclear. It would be nice if the operator $e^{tH}e^{-tL}$ could have a definite correspondence to $e^{itH}e^{-itL}$.

10. The theory as a whole should provide a basis for the study of multiple

stochastic integrals, homogeneous chaoses for processes other than Brownian motion, and thus provide a method for analyzing systematically fairly general functionals of general probabilistic and quantum processes.

BIBLIOGRAPHY

PROBABILITY

Cooper, Hoare and Rahman. "Stochastic Processes and Special Functions: Probabilistic Origin of Some Positive Kernels Associated with Classical Orthogonal Polynomials," J. Math. Analysis and Appl., Vol. 61, No. 1, Nov. 77, p. 262.

Daletskii, Yu. L. "Functional Integrals Connected with Operator Evolution Equations," Russ. Math. Surveys, Vol. 17, No. 5, Sept.-Oct. 62, pp. 1-108.

Feinsilver, P. "Processes with Independent Increments on a Lie Group," Transactions of A.M.S., April 78.

__________ "An Operator Calculus Associated with Processes with Independent Increments," Pacific J. Math., to appear.

Feller, W. An Introduction to Probability Theory and Its Applications, Vol. II, Wiley and Sons, New York, 1971.

Hida, T. Stationary Stochastic Processes, Princeton Univ. Press, 1970.

Kac, M. "On the Average of a Certain Wiener Functional and a Related Theorem in the Calculus of Probability," Transactions of A.M.S., Vol. 59, 1946, p. 401.

_____ "On Some Connections between Probability Theory and Differential and Integral Equations," Proc. 2nd Berkeley Symposium on Math. Stat. and Prob., Univ. of Calif., 1951, p. 189.

Masani, P. "Wiener's Contributions to Generalized Harmonic Analysis, Prediction Theory, and Filter Theory," Bulletin A.M.S., Vol. 72, 1966, p. 73-125.

McKean, H.P. Stochastic Integrals, Academic Press, 1969.

__________ "Geometry of Differential Space," Annals of Probability, Vol. I, No. 2, 1973, pp. 197-206.

Montroll, E.W. "Markoff Chains, Wiener Integrals, and Quantum Theory," Comm. Pure and Appl. Math., Vol. V, 1952, pp. 415-423.

Segal, I. "Non-Linear Functionals of Weak Processes, I," J. Funct. Analysis, 4, 1969, pp. 404-451.

_______ "Non-Linear Functionals of Weak Processes, II," J. Funct. Analysis, 6, 1970, pp. 29-75.

Stroock, D. and Varadhan, S.R.S. "Diffusion Processes with Continuous Coefficients, I," Comm. Pure and Appl. Math., XII, 1969, pp. 345-400.

__________________________ "Diffusion Processes with Continuous Coefficients, II," Ibid., pp. 479-530.

Varadhan, S.R.S. Stochastic Processes, CIMS Lecture Notes, New York University, 1968.

Wasan, M.T. First Passage Time Distribution of Brownian Motion with Positive Drift (Inverse Gaussian Distribution), Queens Univ. Papers in Pure and Appl. Math No. 19, Kingston, Ontario, 1969.

Wiener, N. "Generalized Harmonic Analysis," Acta Mathematica, Vol. 55, 1930, p. 117.

________ Nonlinear Problems in Random Theory, M.I.T. Press and Wiley, 1958.

Wolfe, S.J. "On Moments of Probability Distribution Functions," in Fractional Calculus and Its Applications, Springer-Verlag Lect. Notes in Math., No. 457, pp. 306-316.

MATHEMATICAL PHYSICS

Feynman, R.P. "Space-Time Approach to Non-Relativistic Quantum Mechanics," Revs. Mod. Phys., Vol. 20, 1948, p. 267; reprinted in:

Schwinger, J., ed. Selected Papers on Quantum Electrodynamics, Dover, 1958.

__________ Quantum Electrodynamics, Benjamin, 1962.

Katz, A. Classical Mechanics, Quantum Mechanics, Field Theory, Academic Press, 1965.

Pauli, W. Pauli Lectures on Physics (esp.), Vol. 5, Wave Mechanics, C.P. Enz, ed., M.I.T. Press, 1973.

Simon, B. The $P(\varphi)_2$ Euclidean (Quantum) Field Theory, Princeton Univ. Press, 1974.

________ and Reed, M. Methods of Mathematical Physics, 3 Vols., Academic Press, 1972.

OPERATOR THEORY

Bellman, R. Introduction to Matrix Analysis, McGraw-Hill, 1970.

Davis, H.T. The Theory of Linear Operators, Principia Press, Bloomington, 1936.

Graves, C. "On the Principles which Regulate the Interchange of Symbols," Proc. Royal Irish Academy, 1853-1857, pp. 144-152.

Levitan, B.M. Generalized Translation Operators, Israel Program for Scientific Translations, Jerusalem, 1964.

Rota, G.C. Finite Operator Calculus, Academic Press, 1975.

SPECIAL FUNCTIONS AND ORTHOGONAL EXPANSIONS

Boas, R.P. and Buck, R.C. Polynomial Expansions of Analytic Functions, Springer-Verlag, 1958.

Buchholz, H. The Confluent Hypergeometric Function, Springer-Verlag, 1969.

Haimo, D.T., ed. Orthogonal Expansions and Their Continuous Analogues, Proc. Conf. at S. Illinois U., S.I.U. Press, 1968.

Hochstadt, H. The Functions of Mathematical Physics, Pure and Appl. Math. Series, Vol. XXIII, Wiley, 1971.

Infeld L. and Hull, T.E. "The Factorization Method," Revs. Mod. Phys. 23, 21 (1951).

Lebedev, N.N. Special Functions and Their Applications, R.A. Silverman, Transl., Dover, 1972.

Miller, W. "On Lie Algebras and Some Special Functions of Mathematical Physics," Memoirs of A.M.S., No. 50, 1964.

________ "On the Special Function Theory of Occupation Number Space," Comm. Pure and Appl. Math., XVIII, 1965, pp. 679-696.

Pollaczek, F. "Sur une Généralisation des Polynomes de Legendre," C.R.A.S., Paris, Vol. 228, 1949, p. 1363.

________ "Systemes de Polynomes Biorthogonaux Qui Généralisent les Polynomes Ultrasphériques," Ibid., p. 1998.

________ "Sur une Famille de Polynomes Orthogonaux Qui Contient les Polynomes d'Hermite et de Laguerre comme Cas Limites," C.R.A.S., Paris, Vol. 230, 1950, p. 1563.

________ "Sur une Généralisation des Polynomes de Jacobi," Memorial des Sciences Math., Vol. 131, 1956.

Rahman, M. "A Generalization of Gasper's Kernel for Hahn Polynomials: Application to Pollaczek's Polynomials," Can. J. Math., Vol. XXX, No. 1, Feb. 78, pp. 133-146.

Rainville, E. Special Functions, MacMillan, 1967.

Szëgo, G. Orthogonal Polynomials, A.M.S., Providence, 1967.

Truesdell, C. An Essay Toward a Unified Theory of Special Functions, Ann. of Math. Studies No. 18, Princeton Univ. Press, 1948.

Watson, G.N. Theory of Bessel Functions, Cambridge Univ. Press, 1966.

FURTHER REFERENCES OF INTEREST AND IMPORTANCE

Abramowitz, M. and Stegun, I. Handbook of Mathematical Functions, Nat. Bur. Stds. Appl. Math. Series No. 55, Washington D.C., 1964.

Albeverio, S. and Høegh-Krohn, R.J. Mathematical Theory of Feynman Path Integrals, Springer-Verlag Lect. Notes in Math., No. 523.

Beckner, W. "Inequalities in Fourier Analysis," Ann. Math., Vol. 102, 1975, pp. 159-182.

Howe, R.E. "The Fourier Transform for Nilpotent Locally Compact Groups I," Pacific J. Math., Vol. 73, No. 2, Dec. 77, pp. 307-328.

_________ "On a Connection between Nilpotent Groups and Oscillatory Integrals Associated to Singularities," Ibid., pp. 329-364.

McBride, E. Obtaining Generating Functions, Springer-Verlag, 1971.

Redheffer, R. "Induced Transformations of the Derivative-Vector," Am. Math. Monthly, Vol. 83, No. 4, pp. 255-259.

Roman, S.M. and Rota, G.-C. "The Umbral Calculus," Advances in Math., Vol. 27, No. 2, Feb. 78, pp. 95-188.

Taylor, M. Pseudo-Differential Operators, Springer-Verlag Lect. Notes in Math., No. 416.

Widder, D.V. The Heat Equation, Academic Press, 1975.

Vol. 521: G. Cherlin, Model Theoretic Algebra – Selected Topics. IV, 234 pages. 1976.

Vol. 522: C. O. Bloom and N. D. Kazarinoff, Short Wave Radiation Problems in Inhomogeneous Media: Asymptotic Solutions. V. 104 pages. 1976.

Vol. 523: S. A. Albeverio and R. J. Høegh-Krohn, Mathematical Theory of Feynman Path Integrals. IV, 139 pages. 1976.

Vol. 524: Séminaire Pierre Lelong (Analyse) Année 1974/75. Edité par P. Lelong. V, 222 pages. 1976.

Vol. 525: Structural Stability, the Theory of Catastrophes, and Applications in the Sciences. Proceedings 1975. Edited by P. Hilton. VI, 408 pages. 1976.

Vol. 526: Probability in Banach Spaces. Proceedings 1975. Edited by A. Beck. VI, 290 pages. 1976.

Vol. 527: M. Denker, Ch. Grillenberger, and K. Sigmund, Ergodic Theory on Compact Spaces. IV, 360 pages. 1976.

Vol. 528: J. E. Humphreys, Ordinary and Modular Representations of Chevalley Groups. III, 127 pages. 1976.

Vol. 529: J. Grandell, Doubly Stochastic Poisson Processes. X, 234 pages. 1976.

Vol. 530: S. S. Gelbart, Weil's Representation and the Spectrum of the Metaplectic Group. VII, 140 pages. 1976.

Vol. 531: Y.-C. Wong, The Topology of Uniform Convergence on Order-Bounded Sets. VI, 163 pages. 1976.

Vol. 532: Théorie Ergodique. Proceedings 1973/1974. Edité par J.-P. Conze and M. S. Keane. VIII, 227 pages. 1976.

Vol. 533: F. R. Cohen, T. J. Lada, and J. P. May, The Homology of Iterated Loop Spaces. IX, 490 pages. 1976.

Vol. 534: C. Preston, Random Fields. V, 200 pages. 1976.

Vol. 535: Singularités d'Applications Differentiables. Plans-sur-Bex. 1975. Edité par O. Burlet et F. Ronga. V, 253 pages. 1976.

Vol. 536: W. M. Schmidt, Equations over Finite Fields. An Elementary Approach. IX, 267 pages. 1976.

Vol. 537: Set Theory and Hierarchy Theory. Bierutowice, Poland 1975. A Memorial Tribute to Andrzej Mostowski. Edited by W. Marek, M. Srebrny and A. Zarach. XIII, 345 pages. 1976.

Vol. 538: G. Fischer, Complex Analytic Geometry. VII, 201 pages. 1976.

Vol. 539: A. Badrikian, J. F. C. Kingman et J. Kuelbs, Ecole d'Eté de Probabilités de Saint Flour V-1975. Edité par P.-L. Hennequin. IX, 314 pages. 1976.

Vol. 540: Categorical Topology, Proceedings 1975. Edited by E. Binz and H. Herrlich. XV, 719 pages. 1976.

Vol. 541: Measure Theory, Oberwolfach 1975. Proceedings. Edited by A. Bellow and D. Kölzow. XIV, 430 pages. 1976.

Vol. 542: D. A. Edwards and H. M. Hastings, Čech and Steenrod Homotopy Theories with Applications to Geometric Topology. VII, 296 pages. 1976.

Vol. 543: Nonlinear Operators and the Calculus of Variations, Bruxelles 1975. Edited by J. P. Gossez, E. J. Lami Dozo, J. Mawhin, and L. Waelbroeck, VII, 237 pages. 1976.

Vol. 544: Robert P. Langlands, On the Functional Equations Satisfied by Eisenstein Series. VII, 337 pages. 1976.

Vol. 545: Noncommutative Ring Theory. Kent State 1975. Edited by J. H. Cozzens and F. L. Sandomierski. V, 212 pages. 1976.

Vol. 546: K. Mahler, Lectures on Transcendental Numbers. Edited and Completed by B. Diviš and W. J. Le Veque. XXI, 254 pages. 1976.

Vol. 547: A. Mukherjea and N. A. Tserpes, Measures on Topological Semigroups: Convolution Products and Random Walks. V, 197 pages. 1976.

Vol. 548: D. A. Hejhal, The Selberg Trace Formula for PSL (2, ℝ). Volume I. VI, 516 pages. 1976.

Vol. 549: Brauer Groups, Evanston 1975. Proceedings. Edited by D. Zelinsky. V, 187 pages. 1976.

Vol. 550: Proceedings of the Third Japan – USSR Symposium on Probability Theory. Edited by G. Maruyama and J. V. Prokhorov. VI, 722 pages. 1976.

Vol. 551: Algebraic K-Theory, Evanston 1976. Proceedings. Edited by M. R. Stein. XI, 409 pages. 1976.

Vol. 552: C. G. Gibson, K. Wirthmüller, A. A. du Plessis and E. J. N. Looijenga. Topological Stability of Smooth Mappings. V, 155 pages. 1976.

Vol. 553: M. Petrich, Categories of Algebraic Systems. Vector and Projective Spaces, Semigroups, Rings and Lattices. VIII, 217 pages. 1976.

Vol. 554: J. D. H. Smith, Mal'cev Varieties. VIII, 158 pages. 1976.

Vol. 555: M. Ishida, The Genus Fields of Algebraic Number Fields. VII, 116 pages. 1976.

Vol. 556: Approximation Theory. Bonn 1976. Proceedings. Edited by R. Schaback and K. Scherer. VII, 466 pages. 1976.

Vol. 557: W. Iberkleid and T. Petrie, Smooth S^1 Manifolds. III, 163 pages. 1976.

Vol. 558: B. Weisfeiler, On Construction and Identification of Graphs. XIV, 237 pages. 1976.

Vol. 559: J.-P. Caubet, Le Mouvement Brownien Relativiste. IX, 212 pages. 1976.

Vol. 560: Combinatorial Mathematics, IV, Proceedings 1975. Edited by L. R. A. Casse and W. D. Wallis. VII, 249 pages. 1976.

Vol. 561: Function Theoretic Methods for Partial Differential Equations. Darmstadt 1976. Proceedings. Edited by V. E. Meister, N. Weck and W. L. Wendland. XVIII, 520 pages. 1976.

Vol. 562: R. W. Goodman, Nilpotent Lie Groups: Structure and Applications to Analysis. X, 210 pages. 1976.

Vol. 563: Séminaire de Théorie du Potentiel. Paris, No. 2. Proceedings 1975–1976. Edited by F. Hirsch and G. Mokobodzki. VI, 292 pages. 1976.

Vol. 564: Ordinary and Partial Differential Equations, Dundee 1976. Proceedings. Edited by W. N. Everitt and B. D. Sleeman. XVIII, 551 pages. 1976.

Vol. 565: Turbulence and Navier Stokes Equations. Proceedings 1975. Edited by R. Temam. IX, 194 pages. 1976.

Vol. 566: Empirical Distributions and Processes. Oberwolfach 1976. Proceedings. Edited by P. Gaenssler and P. Révész. VII, 146 pages. 1976.

Vol. 567: Séminaire Bourbaki vol. 1975/76. Exposés 471–488. IV, 303 pages. 1977.

Vol. 568: R. E. Gaines and J. L. Mawhin, Coincidence Degree, and Nonlinear Differential Equations. V, 262 pages. 1977.

Vol. 569: Cohomologie Etale SGA 4½. Séminaire de Géométrie Algébrique du Bois-Marie. Edité par P. Deligne. V, 312 pages. 1977.

Vol. 570: Differential Geometrical Methods in Mathematical Physics, Bonn 1975. Proceedings. Edited by K. Bleuler and A. Reetz. VIII, 576 pages. 1977.

Vol. 571: Constructive Theory of Functions of Several Variables, Oberwolfach 1976. Proceedings. Edited by W. Schempp and K. Zeller. VI. 290 pages. 1977

Vol. 572: Sparse Matrix Techniques, Copenhagen 1976. Edited by V. A. Barker. V, 184 pages. 1977.

Vol. 573: Group Theory, Canberra 1975. Proceedings. Edited by R. A. Bryce, J. Cossey and M. F. Newman. VII, 146 pages. 1977.

Vol. 574: J. Moldestad, Computations in Higher Types. IV, 203 pages. 1977.

Vol. 575: K-Theory and Operator Algebras, Athens, Georgia 1975. Edited by B. B. Morrel and I. M. Singer. VI, 191 pages. 1977.

Vol. 576: V. S. Varadarajan, Harmonic Analysis on Real Reductive Groups. VI, 521 pages. 1977.

Vol. 577: J. P. May, E_∞ Ring Spaces and E_∞ Ring Spectra. IV, 268 pages. 1977.

Vol. 578: Séminaire Pierre Lelong (Analyse) Année 1975/76. Edité par P. Lelong. VI, 327 pages. 1977.

Vol. 579: Combinatoire et Représentation du Groupe Symétrique, Strasbourg 1976. Proceedings 1976. Edité par D. Foata. IV, 339 pages. 1977.